Clinical Applications

for

Anatomy and Physiology

Colleen J. Nolan
St. Mary's University

Kenneth S. Saladin
Georgia College and State University

Boston Burr Ridge, IL Dubuque, IA Madison, WI New York San Francisco St. Louis
Bangkok Bogotá Caracas Kuala Lumpur Lisbon London Madrid Mexico City
Milan Montreal New Delhi Santiago Seoul Singapore Sydney Taipei Toronto

The McGraw-Hill Companies

Clinical Applications for
ANATOMY AND PHYSIOLOGY
COLLEEN J. NOLAN AND KENNETH S. SALADIN

Published by McGraw-Hill Higher Education, an imprint of The McGraw-Hill Companies, Inc., 1221 Avenue of the Americas, New York, NY 10020.

This book is printed on acid-free paper.

10 11 12 CUS CUS 0 9 8

ISBN 978-0-07-285457-2

MHID 0-07-285457-X

www.mhhe.com

Contents

1 Introduction to Clinical Applications

Objectives

In this chapter we will study

- various approaches to the study of disease;
- the role of the Centers for Disease Control and Prevention;
- common causes of disease;
- the distinction between signs and symptoms of disease;
- terms used to describe the time course of a disease; and
- common abbreviations for medical specialists and specialties.

Homeostasis and Disease

The body's tendency to maintain internal stability is called *homeostasis*. Examples include the body's relatively stable temperature, blood glucose concentration, hormone levels, acid-base balance, and electrolyte balance. When physiological variables deviate too much from their *set point,* the body activates *negative feedback loops* that tend to restore stability and maintain health. In some cases, such as the stoppage of bleeding, *positive feedback loops* are activated to bring about rapid change. If the attempt to regain homeostasis fails, *disease* results.

There is a strong emphasis in medicine today on promoting wellness through prevention. However, this manual focuses on what happens when prevention fails, homeostasis is disrupted, and disease occurs.

The Study of Disease

Disease (illness) is any deviation from normal that interferes with correct, life-sustaining bodily function. Literally, the word means *dis-ease,* the opposite of ease (comfort and normal function). Disease may have underlying structural foundations, such as a broken bone, and its effects may be observed not just at the level of bodily form and function but also at the level of the mind, as in psychiatric diseases (mental illness).

The study of disease is called **pathology,** a field that embraces all aspects of disease, from the patient's complaints to the gross and microscopic appearance of dysfunctional tissues and organs. **Pathologists** are physicians and others who specialize in this branch of medicine. A subdivision of pathology called **pathophysiology** focuses specifically on the physiological (functional) aspects of organ dysfunction, as opposed to their structural abnormalities. **Histopathology** is the study of diseased organs at the microscopic level.

Epidemiologists are scientists who study the social distribution and spread of diseases, especially to determine their sources and causes and to halt their spread. Since epidemiology is such an important public health concern and epidemiologists play a key role in formulating public health policy, many epidemiologists work at such organizations as the World Health Organization (WHO), U.S. Public Health Service (USPHS), and comparable national health agencies in other countries. One of the premier institutions for epidemiology is the USPHS division called the Centers for Disease Control and Prevention (CDC), headquartered in Atlanta, Georgia. The CDC was originally established in Georgia because of the prevalence of malaria in that region of the United States and the importance of this infectious disease to the U.S. military personnel who trained at bases in the Southeast. While the CDC is primarily concerned with U.S. public health, its epidemiologists work worldwide because people engaged in commerce and travel so easily carry diseases from one country to another. Disease anywhere in the world is a potential threat to public health everywhere in the world.

The Causes of Disease

Etiology, in the strict sense, means the study of the causes of disease; in the broad sense, it also means the cause itself. For example, you may see a statement that some forms of encephalitis have a viral etiology; this means that they are caused by a virus. Diseases for which no cause can be identified are called **idiopathic** diseases, loosely translated as "disease of one's own."

The causes of disease are enormously diverse; they include:

- genetic disorders such as mutated genes or excess or missing chromosomes;
- immune disorders, in which the immune system is either underactive (as in AIDS) or overactive, attacking the body itself (as in **autoimmune diseases** such as asthma, rheumatic fever, and rheumatoid arthritis);
- infectious agents such as viruses, bacteria, fungi, parasitic worms, and so forth (these organisms transmit **infectious diseases** from person to person, in contrast to *nontransmissible diseases* such as Alzheimer disease or arthritis);
- **trauma** (physical injury) from such causes as blows, cuts, heat, cold, radiation, and electrical shock;
- chemical agents such as **poisons** (any substances taken into the body that disturb normal physiology) and **toxins** (poisons of plant or animal origin);
- nutritional imbalances, ranging from vitamin deficiency diseases and eating disorders *(anorexia, bulimia)* to obesity; and
- stress, which can result from other diseases as well as from *psychosocial causes* such as divorce, the death of a loved one, or having to care for a chronically ill family member.

Diseases present at birth are called **congenital diseases** and may result from several of the preceding causes—for example, trisomy-21 (Down syndrome) from a genetic defect, fetal alcohol syndrome from a poison, congenital syphilis from an infectious organism, or congenital heart defects resulting from developmental abnormalities.

Certain conditions and habits are called **risk factors** because they increase a person's probability of contracting a disease. Some of these we can do nothing about: Old age is a risk factor for osteoporosis and rheumatoid arthritis; being of African descent is a risk factor for hypertension and sickle-cell disease; being of eastern European Jewish descent is a risk factor for Tay-Sachs disease; and being of white European descent is a risk factor for cystic fibrosis and phenylketonuria. Other risk factors are avoidable: Smoking is a risk factor for emphysema and lung cancer; suntanning is a risk factor for skin cancer; and careless sexual activity is a risk factor for AIDS and hepatitis. Such disorders are therefore called **preventable diseases.**

The Signs and Symptoms of Disease

When a person seeks treatment for a disease, he or she becomes a **patient**. This word comes from the Latin *patior,* "to suffer." When a person reports to a clinic or physician (other than for a routine examination), he or she usually has a **complaint,** a feeling of "something wrong." The disease, if indeed one exists, reveals itself through characteristic *signs* and *symptoms.*

A **sign** is an objective indication of disease that can be seen by any trained observer and expressed in terms others can verify—for example, a fever, high blood pressure, unevenly dilated pupils, swollen lymph nodes, or a skin lesion. A **symptom** is a subjective feeling of disease that can be known with certainty only by the patient—for example, pain, fatigue, blurry vision, or dizziness. There is no way that another person can directly perceive another person's pain or dizziness, or even know with certainty that they exist and are not imaginary. A physical examination must correlate the symptoms reported by the patient with the signs observed by the examiner or revealed by clinical tests (such as blood and urine tests). This combination of information is then used to make a **diagnosis** (see chapter 2 of this manual).

Signs and symptoms are sometimes collectively called **pathologies.** Over the course of a disease, there are often typical signs and symptoms that run together. From "run together," we get the word **syndrome** to refer to a collection of signs and symptoms and the degenerative processes that characterize a particular disorder—for example, *acquired immunodeficiency syndrome (AIDS), fetal alcohol syndrome,* and *Down syndrome.*

Diseases in Time

Several important terms in pathology refer to the time course of disease or the status of a disease at a particular point in time. **Prevalence** means the number of people in a given population who have a disease at a given moment *(point prevalence)* or in a given time interval *(period prevalence).* **Incidence** means the number of new cases of a disease that appear in a given population over a given period of time. For example, a disease can have a high prevalence but low incidence, suggesting that it has been brought under control. This occurred in the mid-1900s when many children contracted polio from public swimming pools, but the polio vaccine brought the transmission of this disease under control, so that the number of new cases declined sharply. Thus, the prevalence of polio remained high while its incidence declined.

On the other hand, a disease can have a low prevalence but high incidence, suggesting that it is new to the population and may be an emerging threat to public health. Examples include the emergence of AIDS, ebola, and hepatitis C in recent decades. A high incidence of disease indicates an **epidemic,** an occurrence of illness significantly above normal expectations.

Two more terms that refer to the significance of a disease at the population level are *morbidity* and *mortality*. **Morbidity** is a collective term for the incidence or prevalence of a disease in a population—that is, how many people in a given population have the disease or are coming down with it. **Mortality** means the rate of death in a given population from a particular disease. Certain diseases are called **notifiable (reportable) diseases** because physicians and other health-care providers are required by law to report all known cases to the USPHS or to similar agencies in other countries. These are diseases of special public health importance, making it advisable for the government to be aware of their incidence and prevalence. From such data, the CDC compiles a weekly publication, *Morbidity and Mortality Weekly Reports (MMWR),* which reports the incidence, prevalence, and interesting case studies of reportable and other diseases.

The individual patient is naturally concerned about how long a given disease is likely to last. The **onset** of a disease is the time when signs and symptoms first appear. **Duration** is how long the disease lasts. Two terms distinguish the time course of different diseases in the individual—*acute* and *chronic*. An **acute disease** typically has a sudden onset and a duration of less than 3 months. It may involve one or more days of medical attention and restricted activity. Most acute diseases respond well to medical or surgical treatment; many can be treated with nonprescription drugs. Examples of acute diseases include colds, flu, and appendicitis.

Table 1.1 Some Abbreviations for Professional Titles*

D.D.S.	Doctor of Dental Surgery	M.D.	Doctor of Medicine
D.M.D.	Doctor of Dental Medicine	O.D.	Doctor of Optometry
D.O.	Doctor of Osteopathy	O.T.	Occupational Therapist
E.M.T.	Emergency Medical Technician	P.A.	Physician's Assistant
G.P.	General Practitioner	P.T.	Physical Therapist
L.P.N.	Licensed Practical Nurse	R.N.	Registered Nurse

* The abbreviation P.C. often seen after a doctor's name stands for "Professional Corporation." This is not a professional title, but a business title similar to Inc., serving for state licensing and tax purposes (used also by lawyers, accountants, and other self-employed professionals.

Table 1.2 Some Medical Specialties

Specialty	Area of Concern or Practice
Anesthesiology	Physiology, pharmacology, and clinical basis of anesthesia
Bariatrics	Prevention and treatment of obesity
Cardiology	Study of the heart and treatment of its dysfunctions
Dermatology	Study of the skin and treatment of its dysfunctions
Endocrinology	Study of the endocrine system and treatment of its dysfunctions
Endodontics	Treatment of the dental pulp
Epidemiology	Study of the incidence, distribution, and control of disease in a population
Forensic medicine	The application of medical knowledge to legal matters
Family medicine	Treatment of individuals throughout the life span from infancy to old age, with emphasis on the family as a unit
Gastroenterology	Treatment of the stomach, intestines, and associated organs
General dentistry	General treatment and preventive maintenance of the teeth and associated oral tissues
Geriatrics	Care and treatment of the aged
Gerodontology	Treatment and preventive maintenance of the teeth and associated oral tissues in the aged
Gynecology	Treatment of the female reproductive tract
Internal medicine	Nonsurgical treatment of adult diseases, excluding diseases of the skin and nervous system
Neurology	Treatment of disorders of the nervous system
Nuclear medicine	The use of radioisotopes for diagnosis and therapy
Obstetrics	Care of women during pregnancy, childbirth, and the few weeks immediately after childbirth
Oncology	The study and treatment of both benign and malignant (cancerous) tumors
Ophthalmology	Medical treatment of the eye
Optometry	Examination of the eyes, diagnosis of visual disorders, and prescription of lenses and other aids for improving vision
Orthodontics	Prevention and correction of malocclusion (misalignment of the teeth)
Orthopedics	Medical, surgical, and physical treatment of the musculoskeletal system
Otology	Study and treatment of the ear and hearing
Otolaryngology	Treatment of the ear, nose, and throat
Pathology	Study and treatment of disease
Pediatrics	Treatment of children
Pedodontics	Treatment and preventive maintenance of the dental tissues of children
Periodontics	Treatment of the tissues immediately around the teeth
Pharmacology	Study of drug chemistry, actions, and uses
Podiatry	Treatment of diseases, injuries, or defects of the feet
Psychiatry	Diagnosis and treatment of mental disorders
Radiology	Diagnosis and treatment of disease, or medical imaging, with the aid of high-energy radiation
Rheumatology	Treatment of arthritis and other joint and musculoskeletal disorders
Sports medicine	Treatment and preventive care of people engaged in sporting and recreational activities
Surgery	Treatment of disease, injury, or deformity through manipulation or operation
Teratology	The science of fetal deformity
Toxicology	The science of poisons and antidotes
Urology	Treatment of the urinary tract

A **chronic disease** has a slower onset and lasts more than 3 months. Chronic diseases often cause irreversible pathologic changes and permanent alterations of function. They typically require long-term health care. Examples of chronic diseases include diabetes, arthritis, kidney failure, and some forms of cancer. Some chronic diseases are described as **insidious** because they begin with seemingly minor changes that don't cause much immediate concern. Hypertension and glaucoma, for example, can "sneak up on us," causing irreversible damage or even death (such as a fatal stroke) before any signs or symptoms are observed. Some diseases, intermediate between acute and chronic in their course, are called **subacute (rapidly progressive)** diseases.

Medical Personnel and Facilities

Such a wide variety of treatment facilities are available today that in this manual we use the general term *clinic* to refer to all of them, whether hospital, physician's office, walk-in clinic, long-term care facility, acute-care facility, or other site. Likewise, we use the term *clinician* to refer to a variety of health-care providers: physicians, surgeons, physician's assistants, nurses, aides, dentists, physical therapists, occupational therapists, and many others. Table 1.1 lists some abbreviations for clinicians who practice various medical specialties, and table 1.2 lists some of the specialized branches of medicine.

Case Study 1 The Children with Lead Poisoning

A physician working for the U.S. Public Health Service moves to Los Angeles to assume the directorship of an inner-city health-care facility for the disadvantaged. Over a period of time, she notices that an unusually large number of children brought to the clinic are experiencing joint pain, difficulty walking, and excessive salivation. Some have had seizures. Also, many of their parents note that the children's personalities have changed, with normally outgoing children becoming shy and withdrawn.

The physician decides to investigate the situation and contacts the CDC, which sends an epidemiologist to assist her. They obtain more complete medical histories for 15 children, ranging from 6 to 15 years of age, and perform blood and urine tests on each. In addition to the signs and symptoms already noted, the children frequently report numbness and tingling in their limbs; they perform poorly on hearing, vision, and intelligence tests; and their laboratory results show reduced red blood cell (RBC) counts and traces of lead in the blood and urine.

All of these children live in the same housing project, play together in the neighborhood, and contribute a little to the family finances by salvaging scrap metal from a closed manufacturing plant nearby and selling it to recyclers. When the epidemiologist inspects the site, he finds paint peeling from the factory walls and dust on the floor composed in large part of pulverized paint chips. Analysis of the paint chips and dust reveals a high lead content. (Lead was commonly used in paint before the 1950s) Lead and other heavy metals are also found in soil samples taken around the factory yard.

Suspicious of lead poisoning, the physician initiates a broader campaign of medical examination. Among children under 16 living in the area, she finds a high prevalence of lead poisoning. Specifically, of the 112 children examined during the course of the study, over 70 show at least some signs and symptoms. By contrast, she finds relatively little evidence of lead poisoning among adults 25 and older, who of course do not play on the factory grounds and most of whom have means of employment other than collecting and recycling scrap metal. The only adult with significant indications of lead poisoning is an elderly woman with *pica,* a compulsive habit of chewing on nonnutritive substances—in this case, the lead foil wrapped around wine bottle corks. Pica is often associated with a dietary iron deficiency and with iron-deficiency anemia. The physician treats the affected patients for lead poisoning, and the CDC enlists the Environmental Protection Agency to demolish the old factory and decontaminate the soil.

Based on this case study and other information in this chapter, answer the following questions.

1. What is the etiology of the mental and physiological signs shown by these patients?
2. What risk factors for lead poisoning can you identify in this case study?
3. Is lead a toxin?
4. Do any of the people in this story exhibit idiopathic lead poisoning? Why or why not?
5. Would you consider lead poisoning a syndrome? Why or why not?
6. At what point in this case does histopathology become relevant?
7. Based on the information presented, does the lead poisoning in this community show a high morbidity? A high mortality? A high prevalence? A high incidence? For each term, answer yes or no, or state that there is insufficient information on which to base an opinion. Explain your answers.
8. Would you consider the lead poisoning in this case an epidemic? Would you consider it an infectious disease? Explain each answer.
9. Identify each of the following as either a sign or a symptom of lead poisoning:
   a. joint pain
   b. difficulty walking
   c. excessive salivation
   d. personality changes
   e. low RBC count
   f. subnormal intelligence
   g. dimness of vision
   h. lead in the urine
10. The elderly woman with pica lives with her daughter's family. Her daughter says she can't get her mother to stop chewing the foil from the wine bottles, and the mother says she likes the metallic feel on her teeth. If you were the physician, what might you suggest to control her lead poisoning?

Selected Clinical Terms

acute disease A disease that has a sudden onset and a relatively short duration (less than 3 months), such as acute appendicitis or rhinitis (a cold).

autoimmune disease Any disease in which the immune system attacks the body's own tissues—for example, rheumatic fever, rheumatoid arthritis, and systemic lupus erythematosus.

chronic disease A disease that has a slow onset and relatively long duration (more than 3 months), such as cancer or emphysema.

disease Any deviation from normal that interferes with correct, life-sustaining bodily function.

epidemiology The study of the social distribution and spread of diseases, especially to determine their sources and causes and to halt their spread.

etiology 1. The study of the causes of disease. 2. The cause of any specific disease.

incidence The number of new cases of a disease that appear in a population over a given time period.

infectious disease Any disease caused by organisms such as bacteria, viruses, fungi, etc., that can be transmitted from one person to another.

morbidity The incidence or prevalence of a disease in a population.

mortality The rate of death from a particular disease in a population.

pathology 1. The study of disease. 2. Collective term for the signs and symptoms of a disease.

poison Any substance taken into the body that disturbs normal physiology.

prevalence The number of people in a population who have a disease at a given moment in time or in a given time interval.

risk factor Any condition or habit that increases the probability of contracting a particular disease, such as age, sex, heredity, smoking, diet, or occupation.

sign An objective indication of disease that can be seen by any trained observer, such as a fever or skin lesion.

symptom A subjective indication of disease that can be known with certainty only by the affected person, such as blurry vision or headache.

syndrome A collection of signs, symptoms, and degenerative processes that occur together in a particular disease.

toxin A poison produced by a living organism.

trauma Physical injury to the body, such as a cut, blow, or burn.

2 Diagnosing a Disease

Objectives

In this chapter we will study

- the thought process and procedures involved in clinical diagnosis;
- methods used in the physical examination of a patient;
- common diagnostic tests performed in the laboratory;
- common diagnostic tests performed on the living body;
- some signs of disease; and
- some terms and abbreviations for diseases and diagnostic tests.

Medical Diagnosis

In chapter 1 of this manual, we examined some characteristics of disease. Now we turn our attention to some of the ways a clinician **diagnoses** a disease. The word *diagnose* literally means "to know through"—that is, how a clinician can know, through examination of the patient's signs and symptoms, what the underlying cause is. This chapter describes some of the processes and medical terminology associated with diagnosis.

The Chief Complaint

Diseases are sometimes discovered when patients report for routine medical checkups. Some diseases are **asymptomatic**—they do not produce any discomfort. For example, hypertension (high blood pressure) is nicknamed "the silent killer" because it typically does not produce noticeable symptoms until the damage is already extensive. Since a blood pressure measurement is one of the routine procedures in a medical examination, it may reveal hypertension of which the patient is unaware.

Often, however, a patient **presents** him- or herself at a clinic because he or she feels that something is wrong. The patient's primary medical concern is called the **chief complaint (CC).** When a CC is identified, the search is on for the underlying etiology. The clinician then becomes a medical detective, bent on identifying the culprit so that it can be treated.

Diagnosis as Hypothesis Testing

The process of diagnosis is an application of the *hypothetico-deductive* scientific method. The clinician formulates hypotheses ("These signs could be caused by…") and then embarks on a search for evidence that either supports or rules out each hypothesis.

The first few symptoms the patient reports and the first few signs the clinician observes may lend themselves to multiple interpretations—that is, the cause could be any number of things. For example, the patient may complain of leg pain. This could have a musculoskeletal, neurological, or cardiovascular etiology. The clinician's task is to ask the appropriate questions and observe the appropriate signs to support one of these hypotheses and rule out the others. Are the ankles swollen? This could indicate a cardiovascular problem. Do the patient's joints hurt when he or she moves? This could indicate a musculoskeletal disorder. Does the patient feel pain shooting down the back of the leg? This could indicate a neurological disorder. The clinician can then ask further questions during the interview to progressively narrow down the possibilities: Where is the pain? What is it like? How bad is it? How often does it occur? How long does it last? When does it occur? Is it worse at night; when the patient is at work; when he or she climbs stairs? What factors make the pain better or worse? Has the patient tried analgesics or any other remedy for the pain?

Diagnosis, then, is essentially a process of elimination in which more and more evidence is gathered, alternative hypotheses are discarded when inconsistent with the evidence, and eventually a hypothesis emerges that is supported by the preponderance of evidence.

The Process of Diagnosis

The diagnostic process is commonly described as having four main steps, although they overlap somewhat, and steps 3 and 4 are not always necessary.

1. *Medical History*

The clinician begins by obtaining and recording the patient's **medical history.** This includes **identifying data** (name, address, date of birth, sex, family composition, and occupation); the nature of the chief complaint; the symptoms and duration of the disease; medications or other treatments tried so far; a **family history** (for example, whether any of the patient's close relatives have diabetes, heart disease, cancer, or high blood pressure); and a review of the organ systems other than the one that is the focus of the complaint (How is your hearing? Do you have any trouble breathing? How is your digestion? and so on).

An important part of physician training is the development of good interviewing skills. The physician must be a good listener—able to put the patient at ease, ask the right questions, think analytically about the patient's responses, and adapt his or her approach to the age, sex, education, cultural background, and values of the patient.

2. *Physical Examination*

After this talk with the patient, the clinician carries out a physical examination that involves inspecting the patient's body. The physical examination is also called a **gross examination** because it involves only gross anatomy (that which is visible without magnification). There are several aspects to the physical examination:

Vital Signs Before a patient sees the physician, a nurse typically records the patient's height, weight, and **vital signs:** temperature, heart (pulse) rate, respiratory rate, and blood pressure. The normal ranges of these vital signs are given in the Appendix of Normal Values at the end of this manual. This information is useful not only for assessing the patient's body proportions (whether a patient is underweight, overweight, or obese, for example), but also for determining the proper dosage of some medications.

Inspection The clinician makes note of what can be visually observed about the patient. To some extent, this begins during the interview. While taking the history, the clinician can observe the patient's posture, gait, body proportions, and skin condition (color, texture, and moisture). If the patient has a medical complaint, the relevant area of the body is observed especially closely.

Palpation The clinician touches or presses on the body surface with the fingertips or palmar surface of the fingers and hands. This procedure provides information about body surface temperature, turgor (firmness of the skin), texture, moisture, shape, and vibrations. There are specific techniques for properly palpating different regions of the body, such as the neck, abdomen, breasts, prostate gland, and joints.

Percussion The clinician places the distal segment of the middle finger of one hand on an area such as the chest or abdomen and strikes it sharply with the middle fingertip of the other hand. This produces a vibration and sound. Clinicians are trained to recognize five different tones *(percussion notes)*—called flatness, dullness, resonance, hyperresonance, and tympany—that give information about the density of tissues, accumulations of air, fluid, or scar tissue, and other normal and abnormal conditions.

Auscultation The clinician listens, usually with a stethoscope, to sounds produced by the body, especially respiratory and cardiac sounds. Qualities to note include the pitch, intensity, and duration of the sound, as well as any added, abnormal sounds such as wheezes or heart murmurs.

As you can see, a thorough physical examination requires the examiner to use his or her senses of vision, hearing, touch, and smell. In addition, clinicians use a variety of instruments to enhance their powers of observation. Besides the *stethoscope*, used for listening to sounds of the heart, lungs, and abdominal cavity, these include the *sphygmomanometer* for measuring blood pressure; the *otoscope* and *ophthalmoscope* for viewing the inside of the auditory canal and eye, respectively; the *vaginal speculum* for examining the vagina and cervix; the *percussion hammer* for testing reflexes; and a *tuning fork* for producing the pure tone used in tests of hearing.

The physical examination may provide enough information to diagnose some conditions, but in other cases it only narrows the possibilities to a **preliminary diagnosis.** Further tests may be required, which leads us to step 3.

3. *Laboratory Tests*

Because medical diagnosis often requires knowledge of the current and recent physiological activity of the tissues, a very common step in diagnosis is the performance of laboratory tests on fluid or tissue specimens obtained from the patient. It should be noted here that, although the physician may recommend further tests, the patient has the right to refuse to have them done.

No two fluids are as valuable to diagnosis as blood and urine. For example, if a patient complains of insatiable thirst and hunger and frequent urination, he or she seems likely to have diabetes mellitus. However, laboratory blood and urine tests are needed to confirm that diagnosis.

Blood Tests The blood mediates exchanges between nearly all of the body's tissues and the external environment. It has such a central importance that a great range of diseases are diagnosed (or first discovered, as in the case of asymptomatic diseases) with the aid of data obtained from the blood. The diagnostic examination of the blood is called **clinical hematology.** Some blood tests include red cell (RBC), white cell (WBC), and platelet counts; a differential white cell count (relative numbers of the different classes of white blood cells); the red cell fragility test; bleeding and coagulation (clotting) times; blood gas measurements; hormone, enzyme, and cholesterol levels; hemoglobin concentration; ABO and Rh blood typing; and blood urea nitrogen (BUN) level, a measure of how well the kidneys are clearing metabolic wastes from the blood.

A **complete blood count (CBC)** includes the total and differential WBC counts, RBC count, hematocrit (percent of the blood composed of RBCs), hemoglobin concentration, and platelet count. A CBC provides valuable information for diagnosing infections, allergies, inflammation, bone marrow failure, anemia, leukemia, sickle-cell disease, infectious mononucleosis, and many other conditions.

The concentrations of hormones, enzymes, cholesterol, and other blood components are often expressed in terms of quantities in blood serum. **Serum** is produced by separating the blood plasma from the cellular components, letting the plasma coagulate (clot), and then removing the coagulated protein.

Urinalysis In addition to blood work, **urinalysis (U/A)** is an important part of routine examinations and diagnostic sleuthing. Among the properties examined in urinalysis are urine color and odor, pH, specific gravity (a measure of its concentration of solids), and tests for the presence of blood, pus, protein, glucose, sodium chloride, urates, and phosphates. Disease is indicated by the presence of substances that do not normally appear in the urine, such as protein *(proteinuria),* blood *(hematuria),* ketones *(ketonuria),* or pus *(pyuria).*

Other Common Tests Besides blood and urine, other specimens often examined include throat swabs, which are cultured to check for *Streptococcus* ("strep") or other infectious organisms; sputum, the mixture of mucus and other matter that is coughed or spit up; semen, which is examined for sperm count, sperm motility, and sperm morphology in tests for infertility; and feces (stool samples), which are examined for parasites or blood (a possible indication of colon cancer or polyps).

The removal and microscopic examination of living tissue from a patient is called a **biopsy.** A relatively simple and common biopsy is the **Pap smear,** in which cells are scraped from the uterine cervix and examined for signs of cervical cancer. A **shave biopsy (skin scrapings)** may be used to diagnose infestations by fungi, mange mites, or other pathogens. **Needle biopsy,** in which small bits of tissue are withdrawn through a needle, is used to test for muscle diseases and to sample tumors to test for malignancy.

Aspiration is the removal of a body fluid (liquid or air) or tissue (bone marrow) by suction through a puncture in the body wall. It is performed either for diagnostic purposes or to relieve pressure and other pathologic conditions. The suffix *–centesis* denotes a puncture of the body wall made for such purposes as aspiration. Thus, **spinocentesis (lumbar puncture, spinal tap)** is the sampling of cerebrospinal fluid from a membranous sac below the spinal cord; **thoracocentesis** is the aspiration of fluid or air from the pleural cavity; **paracentesis** is the aspiration of fluid from any body cavity, especially the abdomen; and **amniocentesis** is the aspiration of amniotic fluid from the pregnant uterus, usually to check for genetic defects in the fetus.

4. Tests Performed on the Body Itself

Many tests are performed directly on the patient rather than on materials removed from the body. Among these procedures are medical imaging, endoscopy, and electrophysiological techniques.

Medical Imaging In the past, it was common to perform **exploratory surgery**—to open the body and take a look inside to determine what was wrong with a patient. Now, however, this practice has become less common with the invention of several methods of *noninvasive imaging,* techniques for obtaining an image of the inside of the body without having to open a body cavity. Other noninvasive methods of imaging are also available, many of them specialized for certain branches of medicine, such as cardiology.

Endoscopy Endoscopy is the viewing of the interior of a hollow organ or canal with an **endoscope.** The endoscope is a narrow instrument with a light at its tip to illuminate the organ. It may be designed for direct visual examination of the passage, or it may have an optic fiber connected to a video camera and monitor outside the patient. A *bronchoscope* is an endoscope for examining the bronchial tree of the lungs; a *sigmoidoscope* is used for inspecting the lower colon; an *arthroscope* is used for examining joint cavities; and a *laparoscope* is an endoscope for viewing the abdominal cavity. In addition to simply viewing the interior of an organ, endoscopy is used for biopsies and removing foreign bodies. A number of surgical procedures are now performed by endoscopy, ranging from the repair of knee injuries to appendectomies. Endoscopic surgery requires only a few small incisions for the insertion of the endoscope and surgical instruments.

Electrophysiological Techniques Nervous and muscular tissues generate electrical currents through their actions. These currents are conducted to the skin surface, where they can be picked up by metal plate electrodes placed on the skin and connected to a recording device. Three common electrophysiological recording methods are **electrocardiography (ECG),** the recording of the electrical activity of the heart; **electroencephalography (EEG),** the recording of electrical brain waves; and **electromyography (EMG),** the recording of electrical activity associated with muscle contraction.

Final Diagnosis

When sufficient information has been collected, the ideal result is a **final diagnosis** of the apparent cause of the disease. When the preliminary diagnosis indicates that two or more similar diseases may be responsible for a patient's condition, and the choice between them is reached through further testing, the final diagnosis is called a **differential diagnosis.** From the final diagnosis, the clinician can offer a **prognosis** (the predicted course and outcome of the disease) and formulate a plan of treatment.

Terminology of Clinical Conditions, Signs, and Symptoms

The health professions, like most others, have a specialized vocabulary. Not only must physicians, nurses, therapists, and other health-care providers be familiar with the common terms and abbreviations used in medicine, but so must people in other professions, such as medical transcriptionists, attorneys, and health insurance claims analysts. Table 2.1 lists some common word roots, prefixes, and suffixes that denote clinical conditions and their signs and symptoms. Becoming familiar with the meanings of these word parts will help you not only to remember them but also to figure out the meanings of other similar words that you encounter in the future. Table 2.2 lists common abbreviations for clinical conditions, and table 2.3 lists some abbreviations used in patient records.

Table 2.1 Word Elements Denoting Clinical Conditions, Signs, and Symptoms

Root, Prefix, or Suffix	Meaning	Example and Definition
-agra	Intense pain	Pellagra (severe vitamin C deficiency)
-algesia	Pain	Analgesia (relief from pain)
-algia	Pain, ache	Neuralgia (nerve pain)
-asthenia	Weakness	Myasthenia (muscle weakness)
brady-	Slow	Bradycardia (abnormally slow heartbeat)
carcino-	Cancer	Carcinoma (malignant tumor)
-cele	Hernia	Omphalocele (herniated umbilicus)
-cyesis	Pregnancy	Salpingocyesis (tubal pregnancy)
-dynia	Pain, ache	Odontodynia (toothache)
dys-	Abnormal, bad, difficult	Dyspnea (difficulty breathing)
-ectasis	Dilation, stretching	Bronchiectasis (dilation of the bronchi)
-emesis	Vomiting	Hematemesis (vomiting blood)
febri-	Fever	Febrile (feverish)
hyper-	Above normal	Hypertension (high blood pressure)
hypo-	Below normal	Hyponatremia (blood sodium deficiency)
-ia	Condition	Arrhythmia (irregular heart rhythm)
-iasis	Presence of	Cholelithiasis (gallstones)
-ism	Process, condition	Hyperthyroidism (overactive thyroid gland)
-itis	Inflammation	Cystitis (bladder inflammation)
-lepsy	Seizure	Narcolepsy (a sleep disorder)
-malacia	Abnormal softening	Osteomalacia (softening of bones)
-megaly	Abnormal enlargement	Hepatomegaly (enlarged liver)
-oma	Tumor	Osteosarcoma (bone tumor)
onco-	Tumor	Oncogenic (producing tumors)
-osis	Condition	Thrombosis (abnormal blood clotting)
-osis	Increase	Leukocytosis (elevated white blood cell count)
-paresis	Slight paralysis	Hemiparesis (partial paralysis on one side)
-pathy	Disease	Neuropathy (disease of the nervous system)
-penia	Deficiency	Thrombocytopenia (platelet deficiency)
-phobia	Morbid fear	Acrophobia (abnormal fear of heights)
-phoria	Feeling, mental state	Euphoria (feeling unusually good)
-plasia	Formation	Neoplasia (growth of new tissue)
-plegia	Paralysis	Quadriplegia (paralysis of all four limbs)
-porosis	Becoming porous	Osteoporosis (loss of bone tissue)
-ptosis	Sagging, prolapse	Hysteroptosis (prolapse of uterus into vagina)
-ptysis	Spitting up	Hemoptysis (spitting up blood)
pyro-	Fever	Antipyretic (fever-reducing drug)
-rhage	Excessive discharge	Hemorrhage (excessive bleeding)
-rhea	Discharge, flow	Amenorrhea (cessation of menstrual periods)
-rhexis	Rupture, split	Cystorrhexis (ruptured urinary bladder)
-sclerosis	Hardening	Arteriosclerosis (hardening of the arteries)
sepso-	Infection	Sepsis (infection)
septico-	Infection	Septicemia (bacteria in the blood)
-spasm	Twitch, cramp	Phrenospasm (spasms of the diaphragm—hiccups)
-stenosis	Narrowing	Mitral stenosis (narrowing of heart valve)
tachy-	Fast	Tachypnea (abnormally rapid breathing)
terato-	Monster, birth defect	Teratogenic (producing birth defects)
-tocia	Childbirth, labor	Oxytocin (labor-inducing hormone)

Table 2.2 Abbreviations for Clinical Conditions

AD	Alzheimer disease	HVD	Hypertensive vascular disease
AIDS	Acquired immunodeficiency syndrome	MD	Muscular dystrophy
ASHD	Atherosclerotic heart disease	MI	Myocardial infarction
CA	Cancer	MS	Multiple sclerosis
CF	Cystic fibrosis	OA	Osteoarthritis
CHF	Congestive heart failure	PID	Pelvic inflammatory disease
COPD	Chronic obstructive pulmonary diseases	PKU	Phenylketonuria
CVA	Cerebrovascular accident (stroke)	PMF	Progressive massive fibrosis (of lungs)
DM	Diabetes mellitus	RA	Rheumatoid arthritis
DT	Delirium tremens	SIDS	Sudden infant death syndrome
FUO	Fever of undetermined origin	SLE	Systemic lupus erythematosus
Fx	Fracture	SOB	Shortness of breath
GC	Gonococcus (gonorrhea)	STD	Sexually transmitted disease
GI	Gastrointestinal (conditions)	TB	Tuberculosis
GU	Genitourinary (conditions)	TIA	Transient ischemic attack
HDN	Hemolytic disease of the newborn	URI	Upper respiratory infection
HIV	Human immunodeficiency virus	UTI	Urinary tract infection

Table 2.3 Abbreviations Used in Medical Records

A&P	Auscultation and percussion	Hx	Medical history
BMR	Basal metabolic rate	I&O	Intake and output
BP	Blood pressure	IVP	Intravenous pyelogram
BUN	Blood urea nitrogen	LMP	Last menstrual period
Bx	Biopsy	NPO	Nothing by mouth *(nil per os)*
CBC	Complete blood count	OV	Office visit
CPR	Cardiopulmonary resuscitation	Para 1, 2, 3	Number of live births
CXR	Chest X ray	PE	Physical examination
D&C	Dilation and curettage	p/o	Postoperative
DOA	Dead on arrival	pre-op	Preoperative
DOB	Date of birth	pt	Patient
Dx	Diagnosis	ROM	Range of motion of a joint
FBS	Fasting blood sugar	Rx	Treatment, therapy
FH	Family history	stat.	Immediately *(statim)*
GB	Gallbladder	Sx	Symptoms
GI	Gastrointestinal (conditions)	T	Temperature
GTT	Glucose tolerance test	T&A	Tonsillectomy and adenoidectomy
GU	Genitourinary (conditions)	Tx	Treatment
Hb, Hgb	Hemoglobin	U/A	Urinalysis
HCT	Hematocrit	VS	Vital signs
h/o	History of	y/o	Year old

Case Study 2 The Singer with a Sore Throat

Ellen, a 19-year-old college sophomore, arrives back at her dorm room after taking her last semester exam, but she's in no mood for celebrating. She's worried that she got a C on the test, or maybe even worse, because she is not feeling well. Her throat is so sore that she can hardly talk, and it hurts to swallow. She also feels exhausted and has a headache, although she blames those conditions on the fact that she has stayed up late every night the past week studying for finals. She's also been kept awake by the incessant coughing of her roommate.

Ellen longs to be home, but she has a 3-hour drive ahead of her and hasn't even begun to pack. She skips lunch and sips a can of soda while throwing her dirty laundry and other belongings into assorted boxes, bags, and suitcases. Loading up requires several trips to her car, which is parked in a 1-hour zone a block away. During one trek, Ellen realizes that it is starting to snow, and this puts her in an even worse mood. She has to get home as soon as possible because she has promised to sing at her cousin's wedding tomorrow. Slippery roads will really top off her day.

During the drive home, Ellen thinks the heater in her car must be broken because she can't seem to get warmed up. When she arrives at last, her mother takes one look at her flushed face and watery eyes and settles her on the couch with a cup of hot tea and a comforter. She takes Ellen's temperature, which registers 102.2°F. In view of tomorrow's plans, this is definitely not good news. Ellen and her mother decide to make a quick trip to the acute-care facility in hopes of getting a diagnosis and some medication that will help Ellen feel well enough to sing the next day.

Based on this case study and other information in this chapter, answer the following questions.

1. List Ellen's *symptoms.* What specific *signs* indicate that she is ill?
2. What is Ellen's chief complaint?
3. Assume that you are the clinician responsible for obtaining Ellen's medical history. In addition to identifying data, formulate at least three questions that you would ask to help diagnose her disease.
4. What risk factors in Ellen's recent history may have contributed to her illness?
5. Since Ellen is young and her general health appears good, will a sphygmomanometer be needed during the examination?
6. After looking into Ellen's throat, the acute-care physician feels both sides of her neck. Which physical examination technique is he applying?
7. The physician listens to Ellen's heartbeat and then applies a stethoscope to various points on her back, asking her to take deep breaths and release them as he does so. This is an example of ________________.
8. To look inside Ellen's ears, the physician uses a/an ___________________.
9. The physician uses a long swab to remove a sample from the back of Ellen's throat. What is the purpose of this?
10. Ellen's lab test is negative for *Streptococcus* bacterium, and the physician tells her she probably has a viral infection that will just have to run its course. Ellen takes acetaminophen for her fever and manages to make it through her cousin's wedding. After resting at home for several days, her sore throat goes away. However, she continues to feel run-down, doesn't have much appetite, and spends most of her semester break sleeping. Before going back to school, she decides to visit her regular physician with this latest set of symptoms. Based on the information available, what action would you expect the physician to take at this point?

Selected Clinical Terms

aspiration The removal of tissue, fluid, or air from the body by suction, either to relieve a pathologic state (such as fluid pressure) or to obtain samples for diagnostic examination.

asymptomatic Lacking symptoms even when disease is present.

biopsy The removal and microscopic examination of a sample of living tissue for diagnostic purposes.

chief complaint (CC) The primary reason a patient presents him- or herself for examination or treatment; the major abnormality of structure or function noticed by the patient.

clinical hematology Diagnostic examination of the blood to help assess a person's health or diagnose an illness.

diagnosis Identification of the cause of illness through such methods as the patient interview, physical examination, and laboratory tests.

endoscopy Viewing the interior of the body with a viewing instrument called an endoscope.

medical history A report on a patient that includes identifying data, family medical data, and other information gained through the patient interview.

prognosis Prediction of the course and outcome of a disease as a basis for patient care and treatment.

urinalysis (U/A) Diagnostic examination of the urine to help assess a person's health or diagnose an illness.

vital signs A person's temperature, heart rate, respiratory rate, and blood pressure.

3 Treating a Disease

Objectives

In this chapter we will study

- strategies for the treatment of disease;
- measures for infection control in the clinic;
- medical word roots for clinical procedures;
- types of drugs and routes of administration; and
- symbols and abbreviations used in pharmaceutics.

The Aims of Therapy

Once a disease has been diagnosed, appropriate **therapy (treatment)** can begin. Incurable diseases such as Alzheimer disease can be managed only with **palliative treatment,** which is aimed at alleviating the symptoms and maximizing the patient's comfort rather than curing the disease. **Preventive treatment** is aimed at warding off a disease to which a patient has been exposed or is likely to be exposed. Vaccinations are an example of preventive treatment. If the disease is curable, a plan of **active treatment** is instituted to eliminate the cause and restore homeostasis.

The Therapeutic Partnership

An effective **treatment plan (regimen)** is, at its best, a partnership between the clinician and the patient or the patient's caregivers. Depending on the diagnosis, it may include medicine, surgery, and a combination of diet, exercise, physical therapy, and other restorative activities. Patients have the right to refuse treatment to the extent permitted by law. Some health-care facilities require patients to sign an **informed consent** form testifying that they have been adequately informed of the diagnosis, prognosis, and risks and benefits of the treatment, and have made a voluntary decision to pursue the treatment plan. This is especially important if the treatment plan includes specialized diagnostic procedures, surgery, or experimental therapy. One of the greatest obstacles to successful treatment is **noncompliance,** the failure or inability of the patient or caregivers to follow the plan—for example, by not taking a prescription drug regularly, not adhering to a diet, or not returning for follow-up examinations and care.

Controlling Infections of Clinical Origin

The clinical setting itself poses some challenges to treatment. Many people under treatment contract **nosocomial infections**—infections acquired in a clinic or hospital by exposure to pathogens introduced into that environment by other patients. The very facilities that are intended to cure disease are, for some, the indirect cause of death. Therefore, it is important to rigorously control the spread of pathogens in the clinical environment.

Sterilization means the *complete* removal of microbes from medical instruments and other objects. It is a contradiction in terms to describe something as "partially sterile"—either it is sterile or it isn't. Techniques of sterilization include pressurized steam (as in an *autoclave*), dry heat, ultraviolet light or other radiation, chemicals such as alcohol and phenol, or filtration, as in preparing sterile intravenous solutions. **Disinfection** is the use of chemical and physical agents to destroy microorganisms and their toxins, but it does not completely eliminate them as sterilization does.

Surgical asepsis (sterile technique) consists of procedures that render an area entirely clear of microorganisms and keep it that way. A high level of asepsis is needed in surgery because the patient's body is open and his or her natural defenses against pathogens have been breached. **Medical asepsis (clean technique)** is a lower level of asepsis involving procedures that reduce the population of microorganisms and limit their transfer from person to person, but do not absolutely rid the area of them. Medical asepsis relies partially on the patient's natural defenses, especially the skin and mucous membranes, to provide protection.

Some Treatment Approaches

The approaches to treatment of disease are almost as diverse as the diseases themselves. Some major therapeutic approaches are the following:

- **Chemotherapy**, in the broad sense, is any use of chemicals (drugs) to treat any disease. Originally, it meant the use of drugs that are toxic to a causative organism without being toxic to the patient. The word is now applied especially to the chemical treatment of cancer.
- **Gene therapy** is the replacement of a patient's defective gene with a normal gene. To date, gene therapy is mostly a goal of medical research; it has not yet resulted in a permanent cure for any disease.
- **Hemodialysis** is a procedure that compen-sates for inadequate kidney function by pump-ing the blood through a device that removes its wastes by dialysis.
- **Immunotherapy** attempts to use the body's natural responses to attack cancer cells, other defective cells, or pathogens. Methods include introducing antibodies produced by another individual into the patient's body and boosting the patient's own immune response.
- **Occupational therapy (OT)** is the use of work, play, and self-care activities to prevent or treat physical disabilities and to maximize a patient's independence and quality of life.
- **Physical therapy (PT)** uses exercise, massage, heat, electricity, and hydrotherapy (as opposed to medical or surgical methods) to prevent or treat physical disabilities and maximize a patient's independence and quality of life.
- **Radiotherapy** is the treatment of any disease with ionizing radiation (X rays, ultraviolet rays, or radiation from radioisotopes); it is used especially to destroy malignant tumors.
- **Surgery** is the treatment of diseases, deformities, and injuries by either manipulation or operation (cutting). Surgery without cutting is called *closed surgery;* examples include realigning the ends of a broken bone or treating a dislocated joint.

Terminology of Diagnosis and Treatment

Most medical terminology is composed of word roots that denote an anatomical structure combined with roots, prefixes, or suffixes that denote a normal or abnormal condition or a treatment. Table 2.1 in the previous chapter listed word elements that denote clinical conditions, signs, and symptoms. Table 3.1 lists a number of commonly used word elements that denote diagnostic and therapeutic procedures.

Table 3.1 Word Elements for Diagnostic and Therapeutic Procedures

Element	Meaning	Example
-centesis	Puncture	Amniocentesis (sampling amniotic fluid)
-cision	Cutting	Circumcision (cutting off the foreskin)
-clysis	Irrigation, lavage	Gastroclysis (irrigation of the stomach)
-ectomy	Cutting out	Splenectomy (removal of the spleen)
-graphy	Recording process	Mammography (photographing the breast with X rays)
-iatry	Treatment	Psychiatry (medical treatment of mental disorders)
-metry	Measuring process	Cephalometry (measurement of head dimensions)
-opsy	Viewing procedure	Biopsy (examination of a living tissue sample)
-pexy	Surgical reattachment	Nephropexy (attaching a kidney to the abdominal wall)
-phylaxis	Prevention	Prophylaxis (disease prevention)
-plasty	Surgical repair	Rhinoplasty (repair or remodeling of the nose)
-rhaphy	Suture, sew up	Hysterorrhaphy (suturing the uterus)
-scopy	Viewing procedure	Arthroscopy (viewing the interior of a joint)
-stomy	Making a new opening	Colostomy (making a new opening into the colon)
-therapy	Treatment	Chemotherapy (treatment with chemicals)
-tomy	Making an incision	Phlebotomy (piercing or cutting a vein)
-tripsy	Crushing	Lithotripsy (crushing gallstones or kidney stones)

Basic Facts About Medication

A **drug** is a chemical used to aid in diagnosis, prevent disease, treat disease, relieve pain and other disease symptoms, or improve any physiologic or pathologic condition. Note that this definition focuses on drugs used for therapeutic effect and excludes drugs of abuse. Although food may have some of the same effects that certain drugs have, drugs do not include food; they are nonnutritive chemicals.

Today, a vast number of drugs are available to consumers, both over-the-counter and by prescription only. In the United States, drugs cannot be legally prescribed or sold until they are approved by the Food and Drug Administration (FDA). In order to gain approval, a proposed new drug must pass through a three-phase process of **clinical trials** in which the scientific method is rigorously applied (see Case Study 3 at the end of this chapter).

Routes of Administration

The multiple routes by which drugs can be introduced into the body include oral, transdermal, via mucous membranes, and by injection.

Oral The simplest, most common, and most comfortable route of medication is by mouth. Drugs for oral use are manufactured in the form of **pills** (tiny solid masses containing the drug and diluting and binding substances, which are to be chewed or swallowed), **tablets** (solids compressed into disklike or other shapes), **capsules** (doses of medicine contained in small containers of soluble gelatin), **syrups** (liquid preparations of medicine in a highly concentrated sugar solution), and **elixirs** (medicines dissolved in a sweetened solution of water and alcohol).

Transdermal Many drugs can be absorbed through the skin. Drugs to be administered to the skin are manufactured in the form of lotions (liquids applied externally), liniments (liquids generally applied to the skin with friction), ointments (semisolid, somewhat greasy preparations), and patches (for example, those used to control motion sickness, heart pain, or smoking).

Via Mucous Membranes Certain lotions and liniments are applied to the oral mucosa (the gums, for example), while some pills and tablets (nitroglycerin, for example) are meant to be held under the tongue or between the cheek and gum until the medicine is absorbed. Other medications are administered by **suppository,** a small solid mass shaped for easy insertion into the vagina, rectum, or urethra. The suppository melts at body temperature and releases a drug that is absorbed through the mucous membrane. Drugs may also be given by **enema** or **douche**—flushed in liquid form over the mucous membrane of the rectum or vagina, respectively. Some drugs are given by **inhalation** to be absorbed through the nasal mucosa; nonprescription antihistamines are a common example.

Injection For relatively quick action, some drugs can be injected by needle or catheter into specific tissues: the hypodermis **(subcutaneous,** or **hypodermic, injection);** the muscles **(intramuscular [I.M.] injection);** the veins **(intravenous [I.V.] injection,** or **infusion);** the vertebral canal **(spinal injection);** or the peritoneal cavity **(intraperitoneal [I.P.] injection).**

Types of Drugs

Drugs can also be classified as either *topical* or *systemic*. A **topical** medication is applied externally to a specific region to be treated, such as an ointment used to treat inflammation or skin infections. A **systemic** medication is administered orally, by injection, or in other ways intended to distribute the drug throughout the body. (Some topical medicines become generally distributed after they are absorbed through the skin.)

Still another way to classify drugs is by their intended effect. Table 3.2 lists several classes of drugs defined according to their function. You will encounter these terms again as we discuss specific diseases and their treatments in subsequent chapters of this manual.

Table 3.2 Classification of Drugs by Their Intended Effect

Drug	Effect
Agonist	Produces a physiological action by binding to a receptor for a naturally occurring substance.
Analgesic	Relieves pain without causing unconsciousness.
Anesthetic	Relieves pain by abolishing all sensory perception, either locally *(local anesthetic)* or by producing unconsciousness *(general anesthetic).*
Anthelmintic	Destroys parasitic worms or enables the body to expel them.
Antibiotic	Extracted from molds or bacteria; prevents other microorganisms from multiplying.
Anticoagulant	Inhibits blood clotting.
Antidepressant	Counteracts depression and thus elevates mood.
Antihistamine	Counteracts the effects of histamine and thus relieves conditions such as congestion and allergy.
Anti-inflammatory drug	Inhibits inflammation.
Antipyretic	Reduces fever.
Antitussive	Relieves coughing.
Decongestant	Relieves congestion or swelling of a tissue due to fluid.
Diuretic	Increases urine output; often used to lower blood pressure.
Expectorant	Promotes bronchial secretion and facilitates the expulsion of mucus from the respiratory tract.
Immunosuppressant	Inhibits the immune response; often used to reduce the body's tendency to reject an organ transplant or tissue graft.
Interferon	A glycoprotein that inhibits the replication of viruses and enhances the immune response; used to treat some forms of cancer and viral infections.
Laxative	Mildly stimulates bowel motility and promotes defecation.
Sedative	Inhibits nervous activity or nervous stimulation of a particular organ—for example, a respiratory, gastric, or cardiac sedative.
Tranquilizer	Has a calming or soothing effect without a sedative or depressant effect.

Pharmaceutical Abbreviations

Pharmaceutics is the science of drug preparation, dosage, and administration. A **pharmacist (druggist, apothecary)** has a specialized knowledge of the properties and interactions of drugs and is licensed to prepare and dispense them. Table 3.3 lists some symbols used to specify drug quantities. About 20 drops of water make 1 mL. Three teaspoons make a tablespoon, and 8 teaspoons make a fluid ounce; a fluid dram is 1 tsp. or 1/8 oz. A minim is 1/60 of a fluid dram.

An **international unit (IU)** is a quantity variously defined for different substances—especially fat-soluble vitamins, enzymes, hormones, and vaccines—by the International Conference for the Unification of Formulae. It is defined by its biological effect and therefore varies from one substance to another. For enzymes, an IU is the amount of enzyme that produces 1 μmole of product per minute under standard conditions of temperature, pH, and substrate concentration. For various hormones, an IU is the amount that produces a physiological effect equal to that of a specified quantity (such as 1 μg or 1 mg) of a purified hormone preserved at a specified institution.

Table 3.3 Units of Drug Measurement

Dram	ℨ
Drop, drops	gt, gtt
Gram	g
IU	international unit
Milligram	mg
Milliliter	ml, mL
Minim	m or min
Tablespoon	tbsp.
Teaspoon	tsp.

Case Study 3 Drug Development and Scientific Method

The process of developing a new drug and getting government approval to put it on the market is a long and expensive one, primarily because of the need to protect the public health from unexpected or undesirable side effects. The process begins with animal tests to check for toxicity and progresses into **clinical trials,** which use human subjects. These subjects are paid voluntary participants who have signed informed consent forms. Clinical trials are carried out in three phases that involve increasing expenditures of time and money. The following hypothetical scenario presents some time and cost figures typical at the end of the twentieth century.

Let us suppose that a pharmaceutical company we'll call MedPharm is developing a new analgesic, NoPain, for the treatment of chronic pain in cancer patients. Animal trials reveal no biological toxicity at realistic doses, so with the approval of the Food and Drug Administration, the drug moves into clinical trials.

Phase I clinical trials are meant to screen a drug for safety in healthy human subjects. In the case of NoPain, phase I establishes the maximum safe dose of the drug and shows that NoPain causes no side effects. Even though only 50 healthy volunteers are involved, this phase lasts about a year and a half and costs MedPharm $10 million.

Development moves on to phase II, in which the researchers establish the effect sought for the drug and the optimal dosage regimen (how much should be given, how often, and for how long) to achieve this effect. This phase requires a placebo (control) group of subjects and a randomized double-blind procedure. Phase II testing of NoPain lasts 2 years, involves 300 volunteer cancer patients, and costs MedPharm another $20 million. This phase establishes who will benefit from the drug (perhaps only those with certain forms of cancer), what benefits will be realized, and the best way to administer the drug.

Phase III clinical trials involve a far greater number of patients than phase II—up to 30,000 patients of different age groups in many different clinics—and again includes a placebo group. Extending the trials to a much larger sample size and set of circumstances in this way should establish a much higher degree of statistical confidence in the drug's effectiveness in a broad range of patients. The objective for phase III testing of NoPain is to confirm that the drug is effective in treating people who have cancer-related pain. This phase costs MedPharm another $45 million and lasts for 3 years. At least two trials, involving a statistically significant sample of patients plus control subjects, must show significantly greater effect than is seen in placebo-treated patients. Since NoPain meets this condition, MedPharm files a New Drug Application with the FDA so that the drug can be marketed.

Based on this case study and other information in this chapter, answer the following questions.

1. Phase II of the NoPain clinical trials requires a placebo group, whereas phase I does not. Explain why.
2. Phase I of the NoPain trials uses only 50 subjects. Why couldn't MedPharm save money by limiting phases II and III to 50 subjects each?
3. Would you classify NoPain as a palliative, active, or preventive treatment for cancer? Explain.
4. Mae, a 76-year-old retiree living on a fixed income, complains about the expense of the medications she must take to control her blood pressure and diabetes. "Why do I have to pay as much as $1.33 for one little pill?" she grumbles. Based on your reading of this case study, what might you tell Mae about the cost of doing business from a drug company's point of view?
5. Suppose you were a clinician with plenty of nonsterile cotton swabs on hand, but you needed sterile swabs. Which of the various sterilization techniques might you use to sterilize the swabs? Which of those techniques would not be suitable? Explain.
6. Which of the following treatments may be employed in treating cancer—immunotherapy, chemotherapy, radiotherapy, all of them, or none of them?
7. A patient recovering from surgery has developed a nosocomial lung infection and asks his physician why this occurred. If you were the physician, how would you explain it?
8. For the patient in question 7, the physician prescribes antibiotics. If the physician wants to provide the patient with both rapid and prolonged doses, what route or routes of administration will be used?
9. Why is the use of international units important in medicine?
10. The word root *gastro-* means "stomach." What is the difference between *gastrotomy, gastrectomy, gastroscopy, gastrocentesis,* and *gastroclysis?*

Selected Clinical Terms

chemotherapy Any use of drugs to treat disease, especially the chemical treatment of cancer.

disinfection The incomplete destruction or removal of microbes from medical instruments, surfaces, or treatment areas; sufficient to reduce a patient's risk of infection to an acceptably low level.

drug Any chemical used for the purpose of diagnosis, prevention, or treatment of disease, for the relief of symptoms, or for the improvement of any physiologic or pathologic condition.

gene therapy The replacement of a defective, disease-causing gene with a normal gene.

hemodialysis A procedure that compensates for insufficient kidney function by artificially removing wastes from a patient's blood.

immunotherapy The clinical use of antibodies or immune cells to treat infection, cancer, and other diseases.

informed consent An agreement signed by a patient to receive certain clinical tests or treatments, attesting that the patient has been fully informed of the purposes, risks, and benefits of the proposed procedures and has voluntarily decided to pursue the plan of testing or treatment.

international unit (IU) A quantity of a drug—especially fat-soluble vitamins, enzymes, hormones, and vaccines—defined differently for various substances by the International Conference for the Unification of Formulae. An IU is defined as the amount of a substance that produces a specified physiological effect.

medical asepsis Procedures that reduce the population of microbes in an area to an acceptably low level. Also called *clean technique.*

noncompliance Failure of a patient or his or her caregivers to follow a plan of medical prevention or treatment, often presenting a significant obstacle to successful treatment.

nosocomial infection An infection acquired by exposure to pathogens in a clinical setting.

occupational therapy (OT) The use of work, play, and self-care activities to prevent or treat physical disabilities and to maximize a patient's independence and quality of life.

pharmaceutics The science of drug preparation, dosage, and administration.

pharmacist A person who has a specialized knowledge of the properties and interactions of drugs and who is licensed to prepare and dispense them.

physical therapy (PT) The use of exercise, massage, heat, and other means other than medication or surgery to prevent or treat physical disabilities and to maximize a patient's independence and quality of life.

radiotherapy The treatment of disease, especially cancer, with ionizing radiation.

sterilization The complete destruction or removal of microbes from medical instruments, surfaces, or treatment areas.

surgery The treatment of diseases, deformities, and injuries by manipulation or operation.

surgical asepsis Procedures that render an area entirely free of microbes. Also called *sterile technique.*

therapy Treatment of a patient aimed at relieving suffering, preventing disease, or curing disease.

4 Cellular Form and Function

Objectives

In this chapter we will study

- diseases that result from malfunctions of human organelles;
- the structures of various pathogens; and
- some diseases caused by viruses, bacteria, protozoans, fungi, and parasites.

Organelles and Disease

Each organelle plays a specific role in the overall function of a cell. Thus, a change in the number or functional capacity of organelles can produce cellular dysfunctions, or diseases. In this section, we consider two examples: mitochondrial cytopathies and lysosomal storage diseases.

Mitochondrial Cytopathies

The term **cytopathy** refers to any disorder of a cell or its components; thus, a *mitochondrial cytopathy* is any disease or cellular dysfunction caused by defective mitochondria. Mitochondrial DNA (mtDNA) contains genes for 13 proteins and some RNA molecules (other than the mRNA that codes for those proteins). Mutations in mtDNA can occur in the human egg before conception and thus be passed on to a child. Sperm mitochondria do not survive in the fertilized egg.

The cells most severely affected are those with the highest energy (ATP) demands—neurons, some receptor cells in the eye, and muscle cells—although mitochondrial diseases are not limited to those cells. Some diseases caused by mtDNA mutations are briefly described here:

- **Leber hereditary optic neuropathy** is characterized by loss of vision, with dark areas in the center of the visual field and abnormal color vision. The mean age of onset is 23 years, and for unknown reasons, about three times as many men as women exhibit the disease.
- **MERRF syndrome** (*myoclonic epilepsy and ragged red fibers*) is a syndrome in which the patient exhibits hearing loss, seizures, myoclonus (sudden, shocklike muscle contractions), loss of motor control, intellectual deterioration, and ragged red (slow-twitch) muscle fibers.
- **Pearson marrow-pancreas syndrome** is a disease of infants characterized by anemia and defective pancreatic function.
- **Kearns-Sayre syndrome** includes paralysis of the muscles of eye movement, infiltration of the retina with pigment cells, and sometimes heart block and incoordination resulting from damage to the part of the brain called the cerebellum. It typically begins before age 20 and has a poor prognosis, with most patients dying in their 20s or 30s.

Lysosomal Storage Diseases

Lysosomes are membrane-enclosed packets of enzymes in the cytoplasm. More than 30 diseases have been linked to deficiencies in these stored enzymes—deficiencies that lead to the inability to metabolize certain macromolecules and to the accumulation of those macromolecules in the cell. Thus, these disorders are called **lysosomal storage diseases.** *Pompe disease* and *type II glycogen-storage disease* are defects of glycogen storage, and *Tay-Sachs disease* and *Gaucher disease* are defects of lipid storage.

Pompe disease is a rare condition in which excess glycogen accumulates in skeletal muscle, heart, liver, and nerve cells. It results from the lack of the enzyme *acid maltase,* which is needed to break down glycogen. The symptoms appear soon after birth and include an enlarged tongue, **hypotonia, dysreflexia, cardiomegaly,** and **hepatomegaly** (see "Selected Clinical Terms" at the end of this chapter for definitions). Most patients die before the age of 2 years, but some individuals survive to adulthood and exhibit a less severe disease called **acid maltase deficiency.**

Type II glycogen-storage disease is similar to Pompe disease in that it involves an accumulation of unmetabolized glycogen in the cytoplasm. It especially affects liver, skeletal muscle, and cardiac muscle cells. The accumulated glycogen interferes with the cytoskeleton and intracellular transport, and can lead to fatal weakness in skeletal and cardiac muscle.

Tay-Sachs disease is characterized by the accumulation of a glycolipid called GM_2 (ganglioside) in nervous tissue. It appears in infancy, and most victims die by 3 to 4 years of age.

Gaucher disease also affects the nervous system. It results from the accumulation of glycolipids called *glucocerebrosides* in cell membranes. Three forms of the disease exist. Type I, the most common, is an adult form characterized by bone lesions and **splenomegaly** (enlarged spleen). Type II is a fatal, infantile form characterized by several neurological abnormalities and splenomegaly. Type III occurs at any time during childhood. It combines some features of the adult form with mild neurological dysfunction. Both Gaucher disease and Tay-Sachs disease occur especially among Ashkenazic Jews of eastern European ancestry.

Pathogenic Organisms

A **pathogen** is defined specifically as any disease-producing microorganism, such as a bacterium or fungus, or more broadly as any disease-producing agent, which may include microorganisms but also viruses, multicellular parasites, and chemicals. In this section, we use the term pathogen to refer to viruses, bacteria, protozoans, fungi, and multicellular parasites.

Viruses

Viruses have some properties in common with living organisms—that is, they are composed of protein and nucleic acids. However, they are not cells, and many or most biologists do not consider them to be alive. Viruses have no powers of movement and no metabolism, and they cannot reproduce on their own. Rather, they spread passively through the body, stick to certain cell surfaces, and enter the cells, where they induce the manufacture of more copies of themselves. After this replication process, the host cell is typically destroyed by the emerging viruses, which go on to invade new cells. This destruction of host cells results in disease. Viruses are very small, ranging from 40 nm to 100 nm in size, and they typically exhibit one or more DNA or RNA molecules enclosed in a protein shell called a *capsid.*

When a DNA virus invades a cell, its DNA enters the host cell nucleus and directs the cell to synthesize the components of new viruses. The viruses assemble themselves within the host cell and then bud from the cell surface. In some cases, the viral DNA becomes incorporated into the host DNA.

RNA viruses have two replication mechanisms. In one case, RNA directly stimulates the cell to synthesize more RNA and capsid proteins. These components assemble themselves into new viruses, which are released by lysis (rupture) of the host cell. The other mechanism is seen in **retroviruses,** which contain not only RNA but also an enzyme called *reverse transcriptase.* This enzyme uses the viral RNA as a model to make DNA molecules, which then move into the host nucleus and become incorporated into the host DNA. When these viral genes are active, they induce the host cell to make more viruses. These viruses leave the host cell without lysing it, but their development in the cell disrupts its homeostasis. Descendants of the infected cell may contain viral DNA and also be able to produce more viruses. Retroviruses can remain dormant for years, and then produce disease later when they become activated.

As they emerge from an infected cell, some viruses acquire an *envelope* composed of host cell nuclear envelope or plasma membrane. The envelope is like a disguise; since the viruses are coated with material of human origin, they are not detected as foreign by the immune system.

Diseases caused by viruses include influenza, the common cold, hepatitis, AIDS, chickenpox, cold sores, genital herpes, measles, mumps, rabies, yellow fever, and viral gastroenteritis. Some of these are discussed in later chapters of this manual.

Bacteria

Unlike viruses, bacteria are unquestionably living cells, although they are much different from our human cells. Bacteria are called **prokaryotes** because they do not have a membrane-enclosed nucleus; their DNA, forming a single circular chromosome, lies free in the cytoplasm of the cell. Humans and most other organisms are called **eukaryotes** because their DNA is enclosed in a nucleus (*eu* = true, *karyo* = nucleus). Bacteria lack membrane-bounded organelles such as mitochondria and lysosomes, but they have ribosomes, a cell wall, flagella, and surface extensions called *pili*. Most bacteria are less than 2 mm in size, but some range from 1 mm to 100 mm. The largest of these are barely visible to the naked eye.

Bacteria are often described in terms of their shape. Round bacteria are called **cocci** (sing., *coccus*); rod-shaped bacteria are **bacilli** (sing., *bacillus*); and corkscrew-shaped cells are **spirilla** (sing., *spirillum*). The word root *staphylo-* denotes bacteria that are clumped together, as in *staphylococci* (clusters of round bacterial cells). The root *strepto-* indicates bacteria linked together in chains, as in *streptobacilli,* chains of rodlike cells. *Diplo-* refers to a pair of bacteria joined together, as in *diplococcus.*

Our bodies are heavily populated by bacteria. *Staphylococcus aureus* is abundant on the skin surface and sometimes causes disease. We refer to pathologic infections with this species as "staph infections." *Escherichia coli* is one of the common bacteria in our large intestine. It is normally beneficial to our health, but some strains contracted from public swimming pools, unclean food, and other sources cause serious and potentially fatal infections. However, most bacteria pose no threat to human health, and indeed are necessary to our survival.

Bacterial diseases include tetanus, cholera, bacterial meningitis, tuberculosis, typhoid fever, bacterial pneumonia, leprosy, botulism, anthrax, diphtheria, pertussis, and food poisoning. Many of these are discussed in later chapters of this manual.

Protozoa

Protozoa are unicellular organisms such as the familiar *Paramecium* and *Amoeba.* They are eukaryotic cells ranging from 5 mm to 6.5 cm in size. Like bacteria, most protozoans are harmless and are important to the ecosystem. Some, however, are seriously pathogenic. A few examples are described here:

- *Giardia,* a flagellated protozoan, lives in the large intestine and is transmitted to humans in a cyst form in contaminated food and in drinking water contaminated with sewage. *Giardia* is the world's leading cause of water-borne diarrhea and often breaks out in settings as diverse as luxury hotels, public schools, and refugee camps. Infection with *Giardia* is called **giardiasis.**

- *Plasmodium,* a mosquito-borne protozoan, invades the red blood cells (RBCs) and causes **malaria.** *Plasmodium* destroys the RBCs as it multiplies, invading and destroying more and more of them until it causes severe anemia and, if untreated, death. Malaria is one of the world's leading fatal parasitic diseases.

- *Entamoeba* is an ameba that, in cyst form, may infect a human via contaminated water and food, much like *Giardia. Entamoeba* invades the mucous membrane of the large intestine. It is often nonpathogenic, but for obscure reasons, it sometimes becomes seriously pathogenic, producing severe intestinal pain and bloody diarrhea **(amebic dysentery).** If untreated, this infection can lead to death from dehydration and electrolyte loss.

- *Toxoplasma* is a protozoan whose life cycle typically alternates between cats and rodents. Through such sources as undercooked meat, unpasteurized milk, or exposure to infected cats, humans become accidentally infected with the same stage as the rodent. This infection is called **toxoplasmosis.** *Toxoplasma* causes little harm to adults, but it can cross the placenta and cause serious harm to a fetus, including blindness, **hydrocephalus** (a head enlarged by the accumulation of excess cerebrospinal fluid in the brain), profound retardation, and physical helplessness.

As you can see, parasitic protozoans often invade the body by mouth. *Plasmodium,* however, is one of several examples that require the assistance of a **vector,** typically a biting insect, tick, or other arthropod with mouthparts strong enough to puncture the skin and introduce the pathogen. Many bacterial and viral pathogens also employ vectors to get into the human body. Such diseases are often controlled by controlling the vector. For example, eliminating standing water in malarious areas (such as water found in old tires and cans in trash dumps) deprives mosquitoes of a place to breed and significantly helps reduce the incidence of malaria.

Fungi

Fungi are eukaryotic and may be unicellular (such as yeasts) or multicellular (such as molds and mushrooms). We enjoy fungi in such culinary applications as bread-baking, brewing, and side dishes. In the pharmaceutical industry, we use certain molds to produce penicillin and other antibiotics. Fungi are also important as decomposers because they prevent dead plants and animal carcasses from piling up by breaking down their bodies into nutrients available for new growth.

However, several species of fungi invade the human body and cause pathologies ranging from skin irritation to fatal lung infections. *Candida albicans* is a common fungus that causes skin, vaginal, and oral infections. *Candida* infection, or **candidiasis,** is a common complication of AIDS. Another, often fatal AIDS complication, ***Pneumocystis* pneumonia,** is caused by a lung infection with *Pneumocystis,* until recently misidentified as a protozoan. **Cryptococcosis** involves many organs, especially the brain; infection with *Cryptococcus neoformans* is transmitted through bird feces, such as those left by pigeons and sparrows on apartment building ledges. **Histoplasmosis,** caused by *Histoplasma capsulatum,* is transmitted through the feces of most birds and mammals and affects mainly the human liver, spleen, and lymph nodes. **Athlete's foot (tinea pedis)** is caused by a fungus called *Trichophyton.* Any skin infection with a fungus is called **tinea** (also commonly known as *ringworm,* though it is not caused by a worm). Some varieties of tinea are described in chapter 7 of this manual.

Multicellular Parasites

A **parasite** is any organism that lives in or on the body of another, usually feeding on its host's tissues and causing some harm. The general concept of a parasite includes pathogenic bacteria and fungi, but most often the term connotes a protozoan, worm, or arthropod. In this section, we examine a few multicellular animals that live in the human body as parasites.

Several species of *flatworms* live in humans, including tapeworms, lung flukes, liver flukes, and blood flukes. Blood flukes, or *schistosomes,* are another of the world's leading causes of mortality, though largely limited to the tropics. Infection with blood flukes, called **schistosomiasis,** causes severe and often fatal pathology, mainly because of our own immune reaction to the eggs of the worms.

Roundworms (nematodes) are in a different phylum from flatworms and are somewhat more complex in their structure. Some roundworms that infect humans include hookworms, pinworms, and filarial worms. The last of these are transmitted by blood-sucking insects and sometimes cause **elephantiasis,** a grotesque swelling of the extremities that follows infection of the lymph nodes by the worms.

Parasitic arthropods include ticks, mites, fleas, and lice, which live on the skin surface, burrowed into the skin, or in the hair or clothing. Because of their association with the skin, some of these are discussed in chapter 7 of this manual. In addition to the irritation they cause, many parasitic arthropods are vectors for more deadly viral, bacterial, and protozoan pathogens.

Case Study 4 The Unsuspecting Honeymooners

Jason and Mary take a 2-week honeymoon trip to the Caribbean and spend part of their time hiking on various islands. About 2 days after returning home, they both abruptly begin to feel chills and headaches. Their backs, legs, and joints ache severely, and their eyes hurt when they move them. But since they have been carrying heavy furniture into their new apartment, neither of them is very concerned about these symptoms at first.

However, about 90 minutes after he notices the first symptoms, Jason develops a rapidly rising fever. Mary encourages him to lie down, and soon she feels feverish as well. Since this seems to have no relationship to moving furniture and their fevers are getting rapidly worse, they report to the hospital emergency room. While waiting to see a doctor, both develop rashes on their faces.

The doctor who examines the couple records that they exhibit slightly enlarged spleens, *bradycardia, hypotension,* and *lymphadenopathy* of the *cervical* and *inguinal lymph nodes.* Their other vital signs are normal, and laboratory tests find nothing unusual. The doctor advises them to go home, rest, and drink plenty of fluid. Three days later, Jason and Mary feel better and no longer have any of their earlier symptoms. Both return to work.

But the next day, the symptoms recur and the rash spreads over their entire bodies. Their soles and palms are swollen and bright red. Jason and Mary return to the emergency room, where they again exhibit bradycardia and hypotension. This time, laboratory urinalysis indicates *albuminuria* in both patients. Jason's white blood cell (WBC) count is 2,840/µL and Mary's is 3,600/µL. When the emergency room physician notes from their history that they have recently traveled in the Caribbean, she asks if they were in any mosquito-infested areas. After a little thought, they do remember a remote part of one island where there were a lot of mosquitoes, but say they did not suffer many bites because they were using insect repellent. The physician orders another blood test, which reveals the presence of *Flavivirus*. Based on this result, the physician tells Jason and Mary that she believes they have contracted **dengue** (DEN-gee) **fever**, also known as *breakbone disease* because of the skeletomuscular pain it causes.

Dengue fever is a viral disease transmitted by the bite of some species of the *Aedes* mosquito. Once in the human body, *Flavivirus* invades the lymphatic tissues and reproduces there. The physician advises Jason and Mary to rest for the next 2 to 3 weeks and to avoid taking aspirin in case the dengue fever becomes worse and causes internal bleeding. She advises them to use acetaminophen for headache and joint pain, and tells them to report back immediately if they notice any *petechiae*. She tells them that recuperation requires complete bed rest and that even after their recovery, they will feel tired and lack energy for a time.

Based on this case study and other information in this chapter, answer the following questions.

1. Look up the following terms in a medical dictionary, or elsewhere—*bradycardia, hypotension, lymphadenopathy, cervical lymph nodes, inguinal lymph nodes, albuminuria,* and *petechiae*—and state their meanings.

2. Are Jason and Mary's WBC counts normal or abnormal? Do they exhibit *leukocytosis, leukopenia,* or neither? (You will need to look up these terms.) Are their WBC counts relevant to the diagnosis or not?

3. If you were an epidemiologist charged with controlling dengue fever, what important step(s) might you take? If you were a government employee charged with writing a tourist advisory statement on dengue fever for people traveling in the Caribbean, what precautions would you advise tourists to take?

4. Would the invasion of a cell by viruses increase, decrease, or have no effect on the osmolarity of a cell? Explain your answer.

5. Any cell of the body can be affected by a mitochondrial cytopathy. Why, however, are the cells with the highest metabolic rates the most severely affected?

6. Do you think parasitic worms and arthropods are prokaryotes or eukaryotes? Explain.

7. Explain why children never inherit Kearns-Sayre syndrome from their fathers.

8. You are an epidemiologist charged with investigating and putting a stop to repeated outbreaks of giardiasis in an old, run-down, urban public school. Discuss the hypotheses you would form and the investigations you would undertake.

9. List some risk factors for Gaucher disease, malaria, and toxoplasmosis. Based on the information given in this chapter, which of these risk factors are avoidable and which are not?

Selected Clinical Terms

cardiomegaly Pathological enlargement of the heart.

cytopathy Any disorder of a cell or its components.

dysreflexia An excessive, uncontrollable response of the sympathetic nervous system to stimulation, seen especially in children with spinal cord injuries. Fullness of the bladder or rectum can cause such reactions as hypertension, bradycardia, excessive sweating, respiratory congestion, headache, and facial flushing.

hepatomegaly Pathological enlargement of the liver.

hypotonia Pathological loss of muscle tone, allowing muscles to become flaccid and too easily stretched.

pathogen Any disease-causing agent, including chemicals, radiation, viruses, bacteria, fungi, and parasitic animals; often used in a narrower sense to refer especially to microorganisms that cause disease.

splenomegaly Pathological enlargement of the spleen.

vector An animal, especially a biting arthropod, that transmits pathogenic organisms from one person or animal to another; also used more broadly to include inanimate agents that carry pathogens from host to host, such as wind and water.

5 Genetics and Cellular Function

Objectives

In this chapter we will study

- the characteristics of cancer cells;
- some methods of cancer detection;
- several treatments for cancer;
- how pedigrees represent hereditary patterns in families; and
- how genetic counselors determine the risk of transmitting genetic diseases.

Cancer

Cancer is second only to heart disease as a leading cause of death in the United States; approximately one-fourth of the population develop some form of cancer during their lifetimes, and a diagnosis of cancer is one of the most dreaded pronouncements a person can hear. Cancer is not a single disease, but a group of pathologies with certain properties in common (especially the potential for metastasis).

The first column of table 5.1 ranks the 10 most common forms of cancer in the United States in 1997 according to incidence, while the second column gives the mortality rates for those types. (To refresh your memory of the terms *incidence* and *mortality*, see chapter 1 of this manual.) You can see that different forms of cancer differ greatly in mortality, either because they progress slowly, as in prostate cancer, or because they are highly treatable, as in skin cancer. The top 10 cancers in the United States, from the highest to the lowest mortality rate, are cancer of the lung, colon, breast, prostate, pancreas, blood-forming tissue (leukemias), ovary, stomach, brain and other nervous system, and bladder.

Characteristics of Cancer Cells

The change from a normal cell to a cancerous cell is called **transformation.** Cancerous cells lack the structural characteristics and functions of mature, differentiated cells. Cancerous cells of the lung, for example, cannot carry out normal lung functions. Transformed cells exhibit two heritable characteristics: *anaplasia* and *autonomy*. **Anaplasia** is the lack of differentiation—that is, a lack of specialized form and function. Anaplastic cells resemble embryonic cells rather than mature cells. It is uncertain whether they result from mature cells reverting to embryonic form or from the multiplication of tissue *stem cells* that never differentiated in the first place. **Autonomy** refers to the fact that cancer cells are independent of the normal mechanisms that control the rate of cell division.

Cancer cells typically show enlarged nuclei, abnormally dense DNA, and increased mitotic activity. Their rapid mitosis and lack of histological organization enable tumor cells to invade and destroy neighboring tissues. This invasion is called **progression.** In **metastasis,** cells from the primary tumor (original site) break loose, travel in the blood or lymph, and invade distant organs, where they seed new tumors. Thus, lung and prostate cancers often metastasize to the brain, while colorectal, prostate, and breast cancers often metastasize to the lungs and liver. Some cancer cells also exhibit the following intracellular and surface changes:

1. reduction or modification of surface glycolipids and glycoproteins;
2. loss of the glycoprotein *fibronectin*, resulting in changes in cell shape, cell-cell adhesion, cellular migration, and the location of receptors;
3. disruption of the cytoskeleton, so that it is unable to maintain normal cell shape and cancer cells often become rounded;
4. changes in the nuclear envelope, including a decreased number of nuclear pores, development of projections and pockets, and *blebbing,* the formation of vesicular structures on the envelope; and
5. changes in chromosome structure or number. Structural changes can result when portions of a chromosome are broken off, deleted, or duplicated. Changes in chromosome number occur as entire chromosomes are gained or lost.

Table 5.1 Incidence and Mortality of the 10 Most Common Forms of Cancer in the United States for 1997*

Anatomical Site	1997 Incidence	1997 Mortality
Prostate gland	334,500	41,800
Breast	181,600	44,190
Lung	178,100	160,400
Colon	94,100	46,600
Bladder	54,500	11,700
Skin	40,300	7,300
Rectum	37,100	8,300
Uterus (excluding cervical)	34,900	6,000
Mouth	30,750	8,440
Blood-forming tissue (leukemias)	28,300	21,310

* These types are followed, in descending order of incidence, by cancer of the pancreas, ovary, stomach, nervous system, thyroid, cervix, esophagus, lymph nodes (Hodgkin disease), testis, and bone. (Source: American Cancer Society.)

Cancer Detection and Diagnosis

The early signs and symptoms of cancer may include fatigue, pain, **cachexia** (loss of appetite and extreme weight loss), anemia, increased susceptibility to infection, coughing, abnormal bleeding, or a palpable lump. Tumors are often detected by the patient (as in breast, testicular, and lymph node cancers) or by a physician during a physical exam (as in prostate and colorectal cancers). If a lump is found, it is important to determine whether it is benign or malignant. This can be achieved through biopsy, imaging, and examining body fluids (blood, cerebrospinal fluid, urine) for the presence of chemicals called *tumor markers,* which are produced by cancer cells.

In a biopsy, a portion of the tumor is removed for microscopic examination. This enables the **oncologist** (tumor specialist) to determine if the cells have any of the characteristics of cancer cells. MRI or CT scanning can help determine the size of the tumor, the extent to which it impinges on surrounding tissues and organs, and the extent to which it has metastasized.

Tumor markers include elevated levels of *isoenzymes,* the presence of specific antigens, production of inappropriate hormones, and elevated levels of normal fluid constituents. For example, the presence of prostate-specific antigen (PSA) and acid phosphatase in the blood may indicate prostate cancer. Alkaline phosphatase is a marker for bone cancer. Liver, ovarian, and testicular tumors produce α-fetoprotein. An isoenzyme of lactate dehydrogenase, LDH-1, is a marker for ovarian and testicular cancer.

An oncologist grades and classifies tumors during diagnosis. **Tumor grading** describes the amount of cellular differentiation. There are four grades: A grade I tumor resembles the tissue of origin and has some normal function. Grade II and III tumors show increased cell size and variability of shape, increased rates of mitosis, and less and less resemblance to normal tissue. A grade IV tumor shows no resemblance to normal tissue. **Tumor classification** is based on the cell type, tissue of origin, grade, and anatomic site, as well as on whether the tumor is malignant or benign. Tumors are named for their anatomic site and usually contain the suffix *-oma* (for example, *sarcoma, myoma, lymphoma*).

Cancer Treatment

Once a patient has been diagnosed with cancer, the next step is to treat the disease. The same treatments are not used for all types of cancer. The following approaches can be used separately or in combination:

- **Surgery** is the oldest effective cancer treatment, although its usefulness is largely limited to cancers that have not yet metastasized to distant sites. If the cancer has metastasized to nearby lymph nodes, they are removed along with the tumor.
- In oncology, **chemotherapy** means the use of drugs to kill cancer cells (see chapter 3 of this manual for the broader sense of the term). The aim of chemotherapy is to destroy enough of the

cancer cells that the body's natural defenses can eradicate the rest. Cells that are actively dividing are the most vulnerable to chemotherapy, which is therefore most effective against small, rapidly growing tumors. It is not generally effective against brain cancer because of a *blood-brain barrier* that prevents the drug from getting into the brain tissue.

- **Radiotherapy** is the destruction of tumors by exposure to ionizing radiation. It may kill the tumor cells outright, or it may damage them so that they are more vulnerable to chemotherapy. Chemotherapy and radiotherapy are the most commonly used anticancer strategies. However, they also affect normal tissues that have rapidly dividing cells, such as the intestinal epithelium, hair follicles, bone marrow, and gonads.

- **Immunotherapy** is an effort to attack cancer cells with antibodies and immune system cells. These agents are more selective than chemotherapy and radiotherapy, and thus avoid the side effects of those treatments. Immunotherapy also produces a longer-lasting protection because immune cells are long-lived. Immmunotherapy is a relatively new approach that will surely see further development and broader application.

Just as the treatment varies for each cancer, the goal may differ as well. Obviously, complete **remission,** the disappearance of all clinical evidence of the tumor, should always be the goal. A partial response is a reduction in tumor size by more than 50%. **Disease-free survival** means the time between complete remission and the person's eventual death.

Because of advances in cancer detection and treatment, the survival rates for many different types of cancer have improved since the 1970s. In addition, increased public awareness is leading to earlier detection. The sooner treatment begins, the greater the chance for either complete remission or a prolonged, disease-free survival.

Hereditary Diseases and Genetic Counseling

Many diseases are hereditary; they run in families and are found in some ethnic groups more than in others. For example, cystic fibrosis is especially common among whites, sickle-cell disease among people of African descent, and Tay-Sachs disease

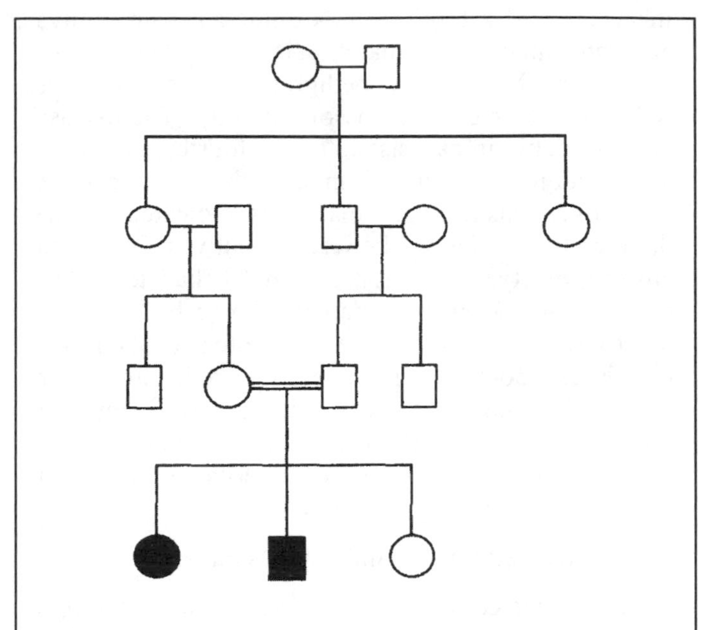

Figure 5.1 Pedigree tracing cystic fibrosis in a given family.

among Ashkenazic Jews. The treatment of such diseases is still almost entirely palliative, aimed at easing suffering but not curing the disease—although ambitious efforts are underway to develop successful gene therapies for hereditary diseases. Treatment of such diseases is the province of the physician, but genetic counselors play an important role in genetic testing, advising couples on the probability of transmitting genetic diseases and helping people cope with the effects of the diseases. One tool of the genetic counselor's trade is a specialized diagram called a *pedigree.*

Interpreting Pedigrees

A **pedigree** shows family relationships and the occurrence of one or more genetic traits among members of the family, usually over three or more generations. By convention, squares on a pedigree represent males, and circles represent females. Horizontal lines stand for matings, and the offspring of any mating are connected to these lines by vertical lines (fig. 5.1). Birth order is indicated by placing the symbols for the offspring in order from left to right. A shaded or darkened symbol indicates the presence of the trait being examined; when only half of the symbol is darkened, it means that the individual is a carrier of the trait. Additional symbols are used to indicate deaths, matings between relatives, and other variables. Pedigrees can be used to determine the specific type of inheritance of a trait, such as dominant or recessive and autosomal or sex-linked.

Figure 5.1 shows a pedigree for cystic fibrosis (CF) over four generations of a family. The disease appeared only in two sisters of the fourth generation. This pedigree is unusual in that the girls' parents were first cousins; their mating is indicated by the double line. Mating between relatives is called **consanguinity.** Each sister with CF had to inherit the CF allele from both parents. Since both parents must have carried the allele, but neither of them had the disease, both of them probably were heterozygous carriers (although it is possible to be homozygous for a trait and still not exhibit it). If a hereditary disease shows the following pattern of inheritance, it is most likely to be *autosomal recessive:*

1. It affects males and females equally.
2. Affected individuals are related through consanguinity.
3. The disease occurs in siblings but not in their parents.
4. On average, it occurs in one-fourth of the children of carrier parents.

Consanguinity is commonly observed in such cases, especially if the trait is rare in the population, because it is more likely that two relatives will share the allele through common descent than that two unrelated people who have the allele will meet by chance and reproduce. For example, about 1 in 25 whites is a heterozygous carrier for CF. The chance of two carriers meeting by chance and reproducing is 1 in 625, and even then, only one-fourth of their children are likely to have CF.

A number of hereditary disorders, including hemophilia and red-green color blindness, are *sex-linked recessive traits*, which show a different pattern of inheritance:

1. The trait appears much more often in males than in females, but may occur in either.
2. The trait is not passed from father to son.
3. The trait can "skip a generation" or more, being transmitted through a series of asymptomatic female carriers.
4. If the father has the trait, all of his daughters receive the allele and are either carriers or exhibit the trait; on average, one-half of these daughters' sons exhibit the trait.

The following case study uses the example of an X-linked recessive trait (hemophilia) to demonstrate how genetic counselors apply a second important tool of their trade—the laws of probability.

Case Study 5 At Risk for Hemophilia

Marcy and Tom are recently married but have not yet had children. They are concerned about the possibility of their children having hemophilia because of a history of the disease in Marcy's family. There is no history of hemophilia in Tom's family. Marcy's doctor refers them to a genetic counselor, who constructs a pedigree of Marcy's family based on his initial interview with the couple (fig. 5.2).

As you can see from the pedigree, Marcy's mother, Florence, was a carrier of the hemophilia gene. Marcy's sister Jessica, the eldest and first in her generation to marry, has two sons with hemophilia, Terry and Greg, and a daughter, Nancy, who is free of the disease. Thus, Jessica is a carrier and can only have gotten the gene from her mother, even though she and her siblings were unaware of any prior history of hemophilia in the family. Neither of Jessica's sisters, Marcy or Danielle, has hemophilia, nor does her brother Richard.

Danielle, however, has married a man with hemophilia, and their daughter Jenny has hemophilia. Danielle was unaware that she was a carrier until Jenny was diagnosed. Now, Marcy and Tom are worried about the risk of their children having the disease. They ask the counselor, "What are the odds?"

The counselor first informs Marcy that, because her mother was a carrier for hemophilia but did not have the disease, there is a 50% probability that Marcy is a carrier. Why? Since her mother had two X chromosomes, one that carried the hemophilia allele and one that did not, there is a 50:50 chance that her mother gave Marcy the defective allele.

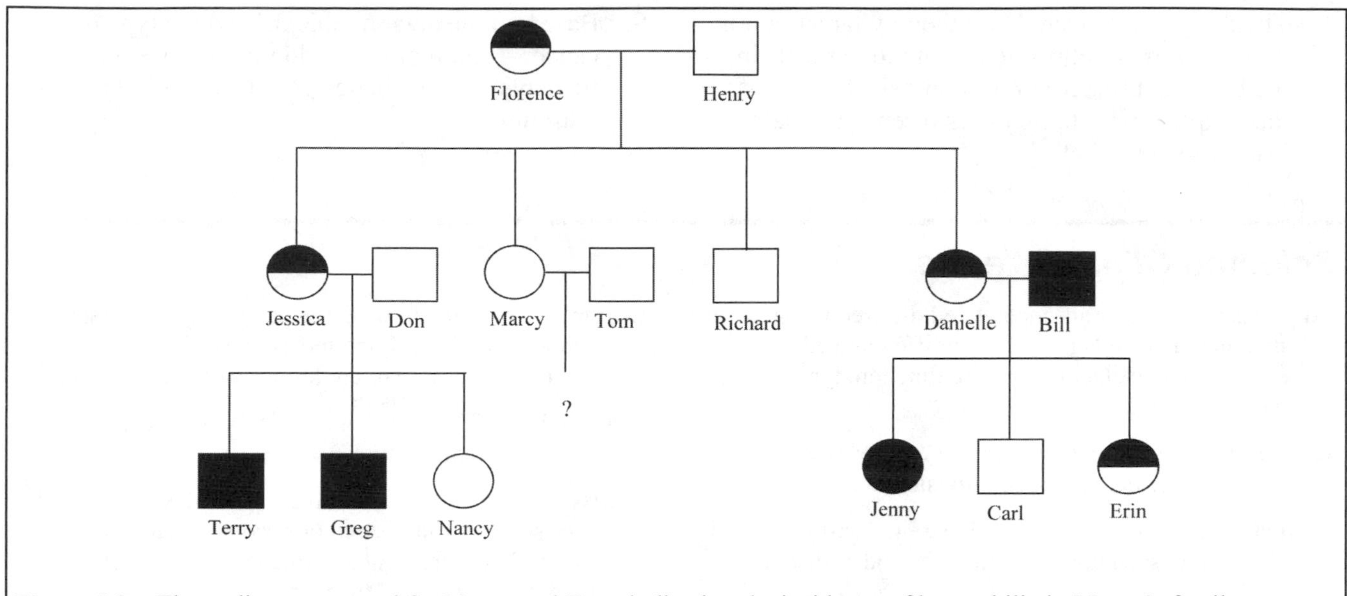

Figure 5.2 The pedigree prepared for Marcy and Tom, indicating the incidence of hemophilia in Marcy's family.

But what is the chance that one of Marcy's children may have hemophilia? For a daughter, the probability is zero. For a girl to have hemophilia, she must inherit the allele from both parents. Tom does not carry the allele; if he did, he would have hemophilia, since men have only one X chromosome and express whatever alleles it carries. Any son of Marcy's, however, could inherit the defective allele from Marcy, and if so, he would have the disease. What is the chance of this?

To calculate this, we use the *rule of multiplication:* If x is the probability of event A happening and y is the probability of event B, then the probability that both A and B will happen is x times y. There is a 50% (0.5) probability that Marcy inherited the hemophilia allele from her mother. If she did, there is also a 50% probability that she will pass it to one of her offspring. The probability of both—receiving it *and* passing it on—is $(0.5)(0.5) = 0.25$ (25%). The genetic counselor can thus advise Marcy that, knowing her mother was a carrier of hemophilia (but not knowing for sure whether Marcy is), there is a 25% chance that any one of Marcy's sons would have hemophilia.

Based on this case study and other information in this chapter, answer the following questions.

1. Review the four principles of sex-linked inheritance listed just before the case study. Which individuals in figure 5.2 demonstrate principle 1?
2. Which individuals in figure 5.2 represent principle 2?
3. Which individuals in figure 5.2 represent principle 3?
4. Why are none of the males in figure 5.2 shown as carriers of hemophilia?
5. What is the probability that Marcy and Tom will have *any* child with hemophilia (irrespective of sex)? Show how you calculate this.
6. Suppose Marcy becomes pregnant and has an amniocentesis, a prenatal cell-sampling technique that can identify the sex of a child (among other things). She finds that she is going to have a baby girl. Should she be worried that the baby will have hemophilia? Why or why not? Suppose she decides to be sterilized after her baby's birth, so she will have only this one child. Should she be concerned that her grandchildren will have hemophilia? Why or why not?
7. Blood typing can help determine paternity. Suppose an unmarried woman gives birth, charges a certain man with being the father, and sues for child support. He demands blood tests, which show that the woman is type A, the baby is type A, and he is type B. Does this evidence rule him out as the father? Explain why or why not.

8. The drugs for cancer chemotherapy target rapidly dividing tumor cells, but also any other cells in the body that undergo rapid mitosis. In light of this, explain why these drugs often cause hair loss as a side effect.

9. Based on the data in table 5.1, which type of cancer—colon or oral—do you think is more treatable (has the better prognosis)? Explain your reasoning.

Selected Clinical Terms

anaplasia A loss or absence of cell differentiation; a state in which cells resemble undifferentiated embryonic cells lacking mature function, typically seen in cancer.

autonomy A state in which cancer cells escape normal controls over the rate of cell division.

cachexia A loss of appetite and extreme wasting away of the body seen in cancer and some other disorders.

consanguinity Relationship through common ancestry ("blood relationship"); marriage or sexual relations between close relatives, increasing the risk of genetic defects in their offspring.

disease-free survival The period of time between complete, permanent remission of disease and a person's eventual death from other causes.

metastasis The spread of cancer from the site of origin to new, remote sites in the body as tumor cells travel in the blood or lymph, lodge in other organs, and seed the development of new tumors.

oncologist A specialist on the properties, cause, pathogenesis, and treatment of tumors, both benign and malignant.

pedigree A chart showing the pattern of relationships in a family and identifying individuals who have exhibited a disease or are known to be carriers of it.

progression The invasion of neighboring healthy tissues by cancer cells.

remission A state in which the signs and symptoms of a disease have disappeared or become notably less severe. May represent a permanent cure or be followed by *relapse,* the reappearance of disease.

transformation The conversion of normal cells to cancerous cells lacking the structure and function of mature, differentiated cells.

tumor classification The categorization of tumors based on cell type, anatomic site, tissue of origin, grade, and whether benign or malignant.

tumor grading The rating of tumor cells according to the degree of cell differentiation.

tumor markers Chemicals detectable in the blood serum or other body fluids that give evidence of a specific type of tumor somewhere in the body.

6 Histopathology

Objectives

In this chapter we will study

- some of the factors leading to cellular injury and necrosis;
- the effects of cell damage or cell death on tissues and organs; and
- how histotechnologists and histopathologists study diseased tissues.

Cellular Injury and Tissue Damage

Cells are vulnerable to injury by such chemical and physical factors as oxygen deprivation *(hypoxia),* free radicals, toxic chemicals, nutritional imbalances, temperature extremes, acids and bases, trauma, and radiation. Cells can often respond to such disturbances by making homeostatic adjustments in their metabolism, but severe disturbances may cause structural changes that can be observed microscopically. The study of such structural changes and the diagnosis of disease from histological evidence—that is, the clinical applica-tion of histology—is called **histopathology.**

Irreversible damage to cell structure and function leads to cell and tissue *necrosis*— pathological death, as compared to *apoptosis,* the "programmed death" of cells that were destined to die in the service of normal bodily development and function. Some of the causes of cellular injury are described here.

Hypoxia

Hypoxia, or oxygen deficiency, is the most common cause of cell death. It can result from **ischemia** (deficient blood flow to a tissue), respiratory disorders such as emphysema that interfere with oxygen pickup by the blood, or various kinds of poisoning that interfere with the blood's ability to transport oxygen (carbon monoxide poisoning) or the tissues' ability to use the oxygen they receive (cyanide poisoning). Hypoxia forces cells to shift to anaerobic fermentation, which may be harmful in two ways: (1) It may produce too little ATP to meet the cell's needs, causing the cell to die from a shortage of usable energy, and (2) hypoxia produces lactic acid, which lowers the pH of the cell, especially in ischemic tissues where blood flow is inadequate to remove the lactic acid from the tissue.

Free Radicals

Free radicals are produced as a normal, toxic by-product of cell metabolism (especially aerobic respiration) and by exposure to radiation and some chemicals. Free radicals can kill cells in three ways: lipid peroxidation (destruction of unsaturated fatty acids), protein fragmentation, and DNA alterations. Of the three, *lipid peroxidation* is probably the most destructive. When the free radical (usually OH•) reacts with the double bond of an unsaturated fatty acid, the reaction generates peroxide. The peroxide then initiates a series of reactions that damage the plasma membrane and the membranes of the cell's organelles, killing the cell. In *protein fragmentation,* peroxide reacts with specific amino acid side chains and irreversibly disrupts the secondary and tertiary structures of the protein, leading to fragmentation. When peroxides interact with many parts of the DNA molecule, *DNA alterations* are induced that destroy the cell's genetic functions. Fortunately, the body has several mechanisms that limit free radical damage.

Chemicals

Chemical injury occurs when a toxic substance damages the various membranes of the cell. This increases membrane permeability and causes the cell and its organelles, especially the mitochondria and endoplasmic reticulum, to swell with water. ATP synthesis and protein synthesis thus break down, gravely compromising the cell's homeostasis. If unchecked, damage to the lysosomes also occurs, and the cell undergoes *autolysis* (self-digestion).

Nutritional Imbalances

Normal cellular function depends on the availability of adequate nutrients such as amino acids, glucose, lipids, minerals, and vitamins. Either excessive or insufficient nutrient levels can harm cells. Usually our diets and the actions of the digestive and cardiovascular systems ensure that cells are appropriately nourished.

Temperature Extremes

Temperature extremes are among the physical factors that can cause cellular injury. Freezing or chilling cells induces *hypothermic injury.* As ice crystals form and melt within cells, they damage the plasma membrane and cause an inflow of sodium ions and water, thus disrupting the cell's osmotic balance. Depending on the rate of chilling, hypothermia can also cause vasoconstriction and ischemia, or it can paralyze the blood vessels in a dilated state, increasing flow and producing severe tissue swelling. In frostbite, blood clotting in the damaged vessels leads to ischemia and gangrene. *Hyperthermia* (excessive body temperature, either because of sun exposure or fever) damages cellular metabolism by speeding up some enzymatic reactions more than others. This causes metabolic pathways to get out of synchrony with each other and leads to derangement of the cell's homeostasis. Extreme heat "cooks" the cell's structural proteins and enzymes, thus causing the immediate tissue death characteristic of burns.

Some Signs and Effects of Cellular Injury

Injured, failing cells are often unable to maintain normal plasma membrane function, so they accumulate fluid. The waterlogged cells have a light-staining, "vacuolated" appearance *(hydropic degeneration).* In addition, failing cells are often unable to metabolize fatty acids, and thus they accumulate lipids in the cytoplasm *(fatty change).* Fatty change is commonly seen in the liver cells of alcoholics, where gross examination (nonmicroscopic inspection) shows a yellowish, greasy-looking **fatty liver.** This can progress to an accumulation of fibrous scar tissue, giving the liver a lumpy surface, a state called **cirrhosis.** Intracellular lipid accumulation is also characteristic of Tay-Sachs and Gaucher diseases (see chapter 4 of this manual).

Other substances, such as pigments, calcium salts, and urate, may also infiltrate damaged cells. Glycogen accumulation is characteristic of Pompe disease and type II glycogen-storage disease (see chapter 4). Pigment accumulation can result from causes as diverse as tattooing, sun exposure, diets high in carotene, or bilirubin accumulation in liver failure.

Dystrophic calcification is the accumulation of calcium salts in dead or dying cells. This is seen in chronic tuberculosis, advanced atherosclerosis, and injured heart valves. Even normal cells accumulate calcium if the concentration of Ca^{2+} in the blood is abnormally high. This is known as **metastatic calcification,** and it can result from vitamin D excess, Addison disease, or bone tumors that decalcify the bones and raise the blood Ca^{2+} level.

Uric acid (or urate, the ionized form) is produced by the breakdown of purines, a component of ATP and nucleic acids. Tissues begin to accumulate crystals of sodium urate if the serum urate concentration rises too high (above 4–5 mg/dL, depending on sex). This triggers an inflammatory condition called **gout,** which may include extremely painful joint inflammation *(gouty arthritis)* and the appearance of small, white nodules *(tophi)* under the skin. Gout can result from abnormal purine metabolism or from insufficient urate excretion by the kidneys. About half of all attacks of gouty arthritis occur in the great toe, and the other half are distributed mainly among the heel, ankle, instep, knee, elbow, and wrist. Gout is at least partially hereditary, and about 95% of its victims are men.

Methods in Histopathology

A **histopathologist** is a physician (M.D.) specialized in recognizing pathological changes in the microscopic appearance of tissues and cells and making diagnoses based on this appearance. A histopathologist is assisted by a staff that may include a pathologist's assistant, histotechnologists, and histotechnicians. A **histotechnologist** is a person who specializes in preparing histological specimens for microscopic examination by the pathologist. A histotechnologist generally must have a baccalaureate degree (preferably in science) and a year of on-the-job training. **Histotechnicians,** who assist histotechnologists, typically have a 2-year diploma or associate degree.

In a hospital setting, histopathology laboratories receive tissues and organs from operating and delivery rooms. Histotechnologists and pathologists often must prepare and examine the tissue and report their findings while the patient is still on the operating table so that the surgeon can determine the appropriate course of action.

Before a tissue can be examined microscopically, it must be sliced into *tissue sections* thin enough to see through—typically about 7 μm thick, which is less than the thickness of many cells. This is done with an instrument called a **microtome,** similar in principle to a butcher's meat slicer but capable of far greater precision. The microtome advances the tissue by fine degrees and shaves off a thin slice each time the tissue passes over the blade of an ultrasharp knife. However, tissues fresh from the body cannot be cut on the microtome because they are too soft; it would be like trying to slice a fresh loaf of bread with a very dull knife. The blade would squash the tissue before it cut it and cause so much distortion in the tissue's structure that the specimen would be useless for diagnostic purposes. Normally, the tissue has to be embedded in a supportive block of paraffin before it is sectioned. A problem with this is that tissue is mostly water, and water and paraffin do not mix. Therefore, some preparation is needed before the embedding process.

The traditional steps of tissue preparation carried out by a histotechnologist are:

- **Fixation** The tissue is cut into small pieces and immersed in formalin or another chemical *fixative,* which prevents decay and somewhat hardens the tissue. Good preservation typically requires at least overnight immersion in the fixative.
- **Dehydration** The specimen is immersed in a series of ethanol baths of increasing concentration, ending with two baths of 100% ethanol, to remove all the water.
- **Clearing** The tissue is treated with a *clearant* such as xylol, clove oil, or wintergreen oil to remove the ethanol. The clearant mixes with paraffin.
- **Embedding** The tissue is immersed in melted paraffin, which infiltrates the tissue and replaces the clearant. The paraffin is then allowed to cool slowly and harden into a solid block.
- **Sectioning** The paraffin block is mounted on the microtome, and a series of very thin slices are cut from it.
- **Mounting** The paraffin-embedded slices are mounted on microscope slides with an adhesive such as albumin. The paraffin is then dissolved out, leaving only a ghostly white or gray tissue section on the slide.
- **Rehydration** In preparation for staining with water-based dyes, the paraffin is dissolved away with a solvent, and the slide is then immersed in a series of progressively lower concentrations of ethanol to rehydrate it.
- **Staining** The slide is immersed in one or more dye solutions that stain different components of the tissue different colors, bringing out the contrast necessary for visual examination. The most commonly used stains are a pair called *hematoxylin and eosin (H&E).* Hematoxylin stains cell nuclei blue to violet, and eosin stains the cytoplasm pink. Different stains are used for specialized purposes such as staining blood, fat, or collagen.
- **Clearing** The stained tissue is treated with a clearing agent again to render it transparent, similar to the way a drop of oil on a piece of paper lets light shine through.
- **Coverslipping** If a specimen is to be kept permanently, a glass coverslip may be applied to the slide with an adhesive. This is not necessary for routine diagnostic work.

Specimen preparation can require several hours to days, because the fixation, clearing, dehydration, infiltration, and rehydration steps must be carried out slowly and carefully to ensure good, useful preparations and to minimize distortion of the tissue. However, if a patient is on the operating table and a surgical decision must be made, there isn't time for this classic technique. The procedure can then be accelerated by freezing the tissue, typically at temperatures of –20°C to –50°C, in a chamber called a **cryostat.** Since freezing hardens the tissue, it takes the place of paraffin embedding—and if paraffin is not going to be used, dehydration, clearing, and rehydration (the most time-consuming steps) are not necessary. The frozen tissue is sectioned with a **freezing microtome** and then thawed and stained.

Once the tissue specimens are prepared, they are examined by a pathologist. In some cases, when specialized methods of diagnosis are needed (such as an assay for estrogen receptors as in the case study that follows), tissue specimens may be sent out to a regional laboratory for preparation and diagnosis. This is routine in the grading of Pap smears, for example.

Case Study 6 Breast Tumors—A Day in the Histopathology Lab

Frances is a board-certified histotechnologist who works at a large urban hospital. Her job consists mainly of receiving tissues and organs from surgeons and preparing them so they can be examined by the pathologist, Dr. Griffin, for diagnostic purposes.

One day, Frances receives specimens from two patients—Ms. Bennett and Ms. Malcolm—who have presented with lumps in their breasts. The oncologist has excised the lumps and sent them to the histology lab. He needs to know whether the lumps are benign or malignant, and if malignant, whether they are local or invasive so that he can decide whether to biopsy the axillary lymph nodes for possible metastatic cancer. The patients remain in the operating room while the specimens are biopsied, so the frozen-section method is used to speed up the diagnostic process.

Ms. Bennett, the first patient, is a single, 28-year-old graduate student. She had not been in the practice of doing breast self-examination (BSE), but was inspired by a newspaper article to start. To her dismay, while doing a BSE in the shower recently, she felt a bean-sized lump in the lower outer quadrant of her right breast. She has visited her physician in deep fear that she has breast cancer.

Dr. Griffin invites Frances to look at the specimen through a dual-head microscope, pointing to a circular, well-circumscribed mass surrounded by a layer of dense, irregular, fibrous connective tissue. Around it are the adipose and areolar tissue typical of normal breast tissue. The mass is composed of loose fibrous tissue and cells that resemble those in the mammary ducts. On the basis of this examination, Dr. Griffin tells Frances that she is diagnosing this as a benign tumor called a **fibroadenoma**, common in young women, and she phones Ms. Bennett's oncologist with the information.

Ms. Malcolm, the other patient, is a 48-year-old attorney and mother of four who has always been in the habit of doing BSEs. She had a baseline mammogram when she was 36 and has been getting mammograms every 2 years throughout her 40s, partly because both her maternal grandmother and her older sister have had breast cancer. Ms. Malcolm has never detected a lump in her own breasts, and yet when she had her last biennial mammogram, a dense mass about 12 mm in diameter appeared in the upper left quadrant of one breast.

Dr. Griffin shows Frances the slides of this biopsy. Mixed among the normal fibrous tissue and adipocytes of the breast are several irregularly shaped clusters of cells and dense fibrous tissue. The cells resemble epithelium, but look immature and are not arranged in epithelial sheets. They are not surrounded by a fibrous capsule, but form tonguelike projections that extend into the surrounding fat and loose fibrous tissue of the breast. Speaking into a voice recorder, Dr. Griffin scans the slide and describes the cells as **pleo-morphic, hyperchromatic,** and anaplastic (see "Selected Clinical Terms" for definitions). Many of the cells that Dr. Griffin points out exhibit **mitotic figures.** A small area of necrotic tissue appears in the center of the lesion. Dr. Griffin, looking a little downcast, says "Well, this one looks malignant—an **invasive ductal carcinoma,** I'd say." She adds that they will send the specimen off to a lab to have a mitotic index and immunoperoxidase assay performed.

Some breast cancers are estrogen-sensitive, and the immunoperoxidase assay is a way of determining whether the malignant cells have estrogen receptors. Immunoperoxidase is a monoclonal antibody that binds to estrogen receptors in cell nuclei. The slides are then washed with a peroxide reagent. If any antibody is bound to nuclear receptors, it reacts with the reagent to produce a brown precipitate. Dr. Griffin shows Frances some older slides so that she can see what a positive assay looks like. Most of the tissue is gray in color, but the cell nuclei are brown. "That means this was an estrogen-sensitive tumor," Dr. Griffin explains. "This is actually good news. It gives Ms. Malcolm a better prognosis, because her cancer may respond to tamoxifen therapy."

Based on this case study and other information in this chapter, answer the following questions.

1. What finding especially suggests that Ms. Bennett's tumor is benign?
2. Why might Ms. Malcolm's doctor recommend biopsy of the axillary lymph nodes?
3. Why would the mitotic index be relevant to a diagnosis of breast cancer?
4. Tamoxifen blocks estrogen receptors so that estrogen cannot bind to them. How could tamoxifen help in the treatment of certain forms of breast cancer?
5. Considering that X rays are known to induce mutations and that mutations can cause cancer, why are women in their 40s and beyond advised to have routine mammograms?
6. Explain why prolonged anaerobic fermentation could cause the enzymes of a cell to become increasingly dysfunctional and eventually lead to a shutdown of the cell's metabolic pathways.
7. What similarity might you expect in the mechanisms of cell injury seen in hypothermic injury and dystrophic calcification? Explain why both disorders may have similar effects on a cell.
8. Mechanical stress or trauma often triggers the symptoms of gouty arthritis in susceptible people. In view of this, explain why gouty arthritis affects the great toe more commonly than other joints.
9. Patients with gout are often advised to drink 3 liters of water daily. Why do you think this would help relieve their symptoms?
10. If you were using conventional paraffin-based histotechnique, but staining the tissue with an alcohol-based stain instead of a water-based stain, what step in the preparation of the slide could you omit?

Selected Clinical Terms

fibroadenoma A benign neoplasm, common in the breast, composed of fibrous connective tissue, proliferating fibroblasts, and anaplastic cells derived from the ductal epithelium of the mammary gland.

histotechnologist A specialist in the preparation of tissue specimens for microscopic examination.

hyperchromatic Staining more intensely than normal with histologic stains.

invasive ductal carcinoma An advanced form of breast cancer in which malignant cells of the mammary ducts have broken through the basement membrane of the duct epithelium and invaded the connective tissue stroma of the breast. This has the worst prognosis of any form of breast cancer.

microtome An instrument that cuts tissue specimens into extremely thin slices suitable for staining and microscopic examination.

mitotic figures Darkly staining aggregates of condensed chromosomes seen in stained cells, indicating that mitosis was underway when the cell was fixed.

mitotic index A count of the percentage of cells in a given area of tissue that exhibit mitotic figures. An abnormally high mitotic index indicates *neoplasia* (the development of a tumor).

pleomorphic Variable in size and shape.

7 The Integumentary System

Objectives

In this chapter we will study

- the importance of the integumentary system in a physical examination;
- terms describing the most common skin lesions;
- some diagnostic tests for skin disorders;
- the rule of nines used to evaluate burn patients;
- pressure ulcers, allergic contact dermatitis, and infections and infestations of the skin and hair;
- diagnostic signs in the nails; and
- acne vulgaris.

Diagnosing Skin Disorders

The skin is the largest, most visible, and most vulnerable of the body organs. Many illnesses cause visible changes in the skin—not only disorders of the skin itself, but also internal disorders such as anemia, lung disease, heart disease, hepatitis, dehydration, and hormone imbalances that affect the skin.

Inspection of the integumentary system is one of the major elements of a comprehensive physical examination. The clinician notes the condition of the facial skin, studies the patient's hands, and palpates the hair and nails, looking especially for abnormal colors, excessive dryness or oiliness, temperature, texture, turgor (how easily a pinch of skin returns to its normal, flat appearance), and the type and distribution of any **lesions** that are present. The clinician may be the first to notice lesions such as skin cancer in areas that the patient cannot see—for example, behind the ears and on the back. When a patient complains of a skin irritation or lesion, the history should include information on potentially harmful substances with which the patient has come in contact—for example, cleaning solvents or poison ivy. Table 7.1 describes many of the most commonly seen skin lesions.

The diagnosis of skin disorders is often complicated by overlapping or nonspecific symptoms such as itching and pain. **Pruritis** (itching) may result from such diverse causes as eczema, infestation with lice, a food allergy, or a systemic disease such as iron deficiency or thyroid trouble. Pain may result from pruritis and the excessive scratching that it stimulates, or it may arise independently from another disease altogether.

Specific skin tests, including the following, can help narrow down the diagnosis:

- **Patch (scratch) test** In this procedure, a known *allergen* (a substance that causes an allergy) is applied to the skin surface or introduced into the skin by scratch or injection. After 1 or 2 days, the site is examined for inflammation (redness and swelling). If inflammation occurs, it indicates that the patient is allergic to that substance.
- **Skin biopsy** A sample of skin is taken (by scraping or shaving the surface) and then microscopically examined. An abnormal histological appearance may be diagnostic of the various forms of skin cancer, among other disorders.
- **Shave biopsy (skin scrapings)** The surface of the skin is scraped and examined microscopically for fungi, mange mites, and other pathogens.
- **Microbial culturing** In this procedure, samples of microorganisms are swabbed from the skin surface and used to inoculate a culture medium. After allowing time for the microbes to multiply, they can be identified by means of biochemical characteristics, nutrient requirements, and microscopic appearance.

Table 7.1 Skin Lesions

Lesion	Description	Examples
	Primary Lesions (may develop from previously normal skin)	
Flat, nonpalpable lesions and changes in color		
Macule	Small (< 1 cm dia.) area different in color from surrounding skin	Freckles, measles, petechiae
Patch	Larger than a macule (> 1 cm dia.), irregular in shape	Vitiligo, port-wine stain
Telangiectasis	Thin, irregular red lines produced by capillary dilation	Rosacea
Palpable, elevated solid masses		
Papule	Small, firm, elevated area (up to 5 mm dia.)	Wart, elevated mole
Plaque	Flat, elevated, rough lesion over 5 mm in diameter; may result from coalescence of papules	Psoriasis, seborrheic keratosis
Nodule	Elevated lesion over 5 mm in diameter; often deeper and firmer than a papule	Lipoma
Tumor	Large (> 2 cm dia.), deeper nodule	Neoplasm, lipoma
Wheal	Elevated, irregular area of temporary cutaneous edema	Insect bites, hives
Superficial elevated lesions with fluid-filled cavities		
Bulla	Elevated vesicle over 5 mm in diameter filled with serous fluid	Blister, pemphigus vulgaris
Cyst	Elevated, encapsulated lesion in dermis or subcutaneous tissue; filled with liquid or semisolid matter	Acne, sebaceous cyst
Pustule	Elevated, superficial, pus-filled lesion	Acne, impetigo
Vesicle	Elevated, superficial lesion up to 5 mm in diameter; filled with serous fluid	Shingles, chickenpox
	Secondary Lesions (develop from changes in primary lesions)	
Loss of surface tissue		
Erosion	Loss of epidermis, usually following rupture of vesicle; moist and glistening but not bleeding	Ruptured blister or chickenpox vesicle
Ulcer	Deeper loss of epidermis and dermis, sometimes with bleeding and scarring	Pressure ulcer, syphilitic chancre
Fissure	Linear crack in the skin	Athlete's foot
Accumulated material on skin surface		
Crust	Dried blood, serum, or pus on skin	Impetigo
Scale	Flake of exfoliated epidermal cells	Dandruff, psoriasis
	Other Lesions	
Excoriation	Abrasion or scratch	Scrapes, scratches, scabies
Keloid	Elevated, irregular scar that progresses beyond area of original injury	Postsurgical keloids
Lichenification	Thickening and roughening with exaggerated visibility of normal skin furrows, resulting from persistent scratching, rubbing, or irritation	Chronic dermatitis
Scar	Fibrous tissue replacing injured skin, not extending beyond the injury	Healed wounds and surgical incisions

Common Skin Disorders

Skin disorders have a wide variety of causes, including trauma, infection, allergy, and cancer. This section also describes pressure ulcers, contact dermatitis, and infection with viruses, bacteria, fungi, and parasites.

Burns and the Rule of Nines

The skin is more exposed to trauma than any other organ, and burns are the most serious and common trauma of the skin as well as the leading cause of accidental death. The treatment of a burn and the prognosis for recovery depend on the extent of the burn. Extent is commonly estimated by means of the **rule of nines,** in which the body surface area (BSA) is divided into regions that are each a fraction (one-half) or a multiple of 9. For adults, the head is about 9% of the BSA, each upper limb is 9%, the trunk is 18%, each lower limb is 18%, and the perineum is 1%. For the head, trunk, and limbs, the percentage is divided evenly between the anterior and posterior surfaces. These values differ for children and are often estimated using a table with data arranged according to the child's age. Burns of irregular shape and distribution can be estimated by comparing them to the palm of the patient's hand, which is about 1% of the body surface area, regardless of age.

According to the rule of nines, if an adult were burned over the face and the anterior surface of the trunk and upper limbs, the burn would cover roughly 22.5% of the body. A burn affecting 20% or more of the body surface is considered a major burn; burns affecting two-thirds or more of the body are usually fatal.

Pressure Ulcers

Pressure ulcers are skin lesions caused by *ischemia* (lack of blood flow) and the resulting necrosis (tissue death) in areas of skin subjected to persistent pressure. They are also known as *bedsores* and *decubitus ulcers,* among other names. Pressure ulcers commonly occur in people who are confined to a bed or wheelchair and are unable to change position for long periods. Pressure ulcers occur especially in areas where bone comes close to the surface and the skin is compressed between the bone and the surface on which the patient rests. Thus, they are often seen in the areas of the coccyx ("tailbone"), hips, ankles, and heels. Unrelieved pressure can lead to necrosis and inflammation. The necrotic lesion may extend well beyond the skin, exposing underlying muscle and bone. Pressure ulcers are more common in elderly patients than in younger ones and more frequently seen in whites than in other people. They are also common in people who suffer decreased sensation and are thus unaware of the pressure and growing lesion; therefore, spinal cord injuries and diabetes mellitus are risk factors for pressure ulcers.

An important goal in patient care is to prevent pressure ulcers from forming. This is accomplished by changing the patient's position frequently, ensuring adequate nutrient and fluid intake, and using beds and wheelchairs that have pressure-reducing surfaces. If pressure ulcers develop, treatment is based on their severity. Superficial ulcers affecting only the upper layer of the skin are treated by covering the ulcer with a flat dressing that will not wrinkle and keeping the ulcer moist. Large pressure ulcers that affect deeper layers of the skin and underlying tissues may require debridement (cutting or scraping away necrotic tissue) and opening the lesion to allow drainage. The wound may be closed with a skin graft or, more commonly, a **myocutaneous flap** (created from the patient's own skin, muscle, and underlying blood vessels near the area of the ulcer).

Allergic Contact Dermatitis

One of the many inflammatory diseases of the skin is **allergic contact dermatitis,** seen in allergies to poison ivy and cosmetics, for example. When an allergen contacts the skin, it may bind to a carrier protein and form a *sensitizing allergen.* Dendritic cells of the skin process the allergen and alert the immune system, which then mounts an immune response to the allergen.

An allergic response is not usually seen until the second exposure to an antigen. Symptoms include erythema, pruritis, swelling, and vesicular lesions at the site of contact. Patch testing is most often used to identify the antigen. Removal of the antigen and prevention of further contact are required to stop the inflammatory response and allow for tissue repair. Severe inflammation may require either topical or systemic treatment with a corticosteroid such as hydrocortisone.

Latex allergy is of increasing concern in health care because clinicians are exposed to such a wide variety of products made of latex, including catheters, tubing, elastic, and gloves used for surgery, examinations, and cleaning.

Viral Infections

Many different viruses cause skin diseases. Examples include cold sores, warts, shingles, and chickenpox. **Chickenpox** is a common disease of early childhood caused by the varicella-zoster virus. The virus is spread through airborne droplets and by close personal contact. The virus causes a fever accompanied by a vesicular rash that usually begins on the scalp, face, and trunk and then spreads to the limbs. Vesicles, papules, and macules appear for up to 5 days. The patient is infectious to others from 24 hours before the rash develops to as long as 6 days after. In time, the vesicles rupture and become encrusted. Treatment methods include baths, wet dressings, and antihistamines to relieve itching. Scratching the rash can lead to a secondary infection, which then may require antibiotic treatment. After the chickenpox clears up, the virus may remain dormant in the nervous system and erupt much later in life (usually after age 50) as a long-lasting, sometimes very painful skin disorder called **shingles (herpes zoster).**

Bacterial Infections

Most bacterial infections of the skin are caused by *Staphylococcus aureus* ("staph infections"). Examples include:

- **folliculitis,** infection of the hair follicle;
- **furuncles** or **boils,** folliculitis that has spread to the surrounding dermis;
- **carbuncles,** aggregates of infected hair follicles; and
- **cellulitis,** infection of the dermis and subcutaneous tissue.

Such conditions are usually treated by cleaning the area with soap and water and applying topical antibiotics. Furuncles and carbuncles are often treated with warm compresses to promote drainage. If the infection persists or spreads to other body systems, oral antibiotics are prescribed.

Fungal Infections (Tinea)

Fungi often feed on the protein keratin and therefore infect the hair, nails, and epidermis. **Tinea,** or fungal infection, is classified by location: tinea capitis, on the scalp; tinea cruris, the groin; tinea pedis, the feet; tinea manus, the hands; tinea unguium, the nails; and tinea corporis, the skin excluding the scalp, face, hands, feet, and genitals. Tinea corporis often has a ringlike border and was therefore named *ringworm* in the fifteenth century, although we now recognize that it is not a worm. Symptoms of tinea vary depending on the body region affected, but itching and scales are common to all types. Diagnosis is usually made by examining skin scrapings and identifying the microbes in culture. Treatment typically involves topical and systemic antifungal drugs. Keeping the infected area clean and dry also helps inhibit fungal growth.

Parasitic Infestations

Parasites are organisms that live in contact with another organism, called the *host,* and usually cause some pathology. Although the aforementioned bacteria and fungi can be considered parasites in the broad sense, the word *parasite* more often denotes organisms such as protozoans, worms, and arthropods that live on and at the expense of a host. The presence of skin parasites is called an **infestation,** as opposed to an **infection,** which is the multiplication of pathogens within the body. Some skin pathogens can be classified either way, such as a fungus that is not confined to the skin surface but also penetrates deeply into the skin.

Some parasitic animals invade the human body accidentally. For example, if a cat with hookworms (genus *Ancylostoma*) buries its feces in a child's sandbox, worm larvae hatch from eggs in the feces and may burrow into the skin of children playing in the sandbox. These larvae cause an infestation called **creeping eruption (cutaneous larva migrans).** The worms become disoriented in this unnatural host, crawl about in the skin, and soon die there. As they burrow through the skin, they create undulating trails that become inflamed, itchy, and crusty. Dogs also carry *Ancylostoma.* When allowed to roam on beaches and defecate in the sand, they set the stage for creeping eruption in people who later sit or lie on that
spot even after the fecal pile has disappeared. Creeping eruption itself is not dangerous, but scratching the itchy lesions can lead to more serious secondary bacterial infections.

Lice are small, parasitic, blood-sucking insects. Three species of lice infest humans: the head louse, *Pediculus humanus capitis,* which lives on the scalp and glues its eggs *(nits)* to the hair; the body louse, *Pediculus humanus corporis,* which lives in the clothing and glues its eggs to the fabric; and the pubic louse ("crabs"), *Pthirus pubis,* a more stocky, crab-shaped louse that lives in the pubic hair.

Infestation with lice of any species is called **pediculosis.** Head lice are especially common among school children. Pubic lice are normally transmitted by sexual contact. For the most part, all three species cause itching, which may be intense enough to interfere with sleep and schoolwork, but nothing more serious. The body louse, however, spreads bacterial diseases, including *epidemic typhus,* which has caused monumental epidemics killing millions of people. The body louse, and thus typhus, spreads from person to person mainly in crowded, unsanitary conditions, such as urban slums, prison camps, flophouses, and crowded mental health institutions, and by sharing beds and clothing with infested people.

Mites are arachnids, related to ticks and spiders. One of the most common mite infestations of humans is **scabies,** caused by the mange mite *Sarcoptes scabiei.* Like *Ancylostoma* larvae, *Sarcoptes* burrows through the skin and causes intensely itchy, crusty lesions. Unlike *Ancylostoma,* however, the infestation does not disappear on its own, and *Sarcoptes* is quite at home in humans, where it reproduces on the body. Scabies is treated with topical medications.

Disorders of the Accessory Organs of the Skin

Nails

The accessory organs of the skin also are subject to pathology. Abnormalities of the nails, for example, can be important indicators of both local and systemic diseases. Lung diseases that cause chronic hypoxemia (a deficiency of blood oxygen) lead to **clubbing** of the fingertips. The fingertip becomes bulbous and the nail more convex. Cirrhosis of the liver and non-insulin-dependent diabetes mellitus can cause **Terry's nails,** in which the nails are abnormally white with a distal band of brown. White spots, lines, pits, and grooves in the nails can indicate psoriasis, various systemic illnesses, or just excessive manicuring. Since many nail markings begin at the nail root and move toward the tip as the nail grows, clinicians can often estimate the time of an illness from the position of the marking and the known rate of nail growth (about 0.1 mm/day in the fingernails). For such reasons, inspection of a patient's fingernails is an important part of a physical examination.

Acne

The most common disorder of the cutaneous glands is **acne vulgaris** (*vulgaris* means "common"). Acne can occur at any age, but is most common and pronounced in adolescence, when the body is adapting to the elevated level of testosterone and other sex steroids. (Testosterone is secreted in both males and females.) Acne affects about 85% of people between 12 and 25 years of age. It is equally common in males and females, but often affects males more severely.

Acne originates in the *sebaceous follicles*—hair follicles with a small vellus hair, large sebaceous glands, and a dilated canal opening onto the skin as a pore. Acne is triggered by a combination of three factors: (1) *hyperkeratosis,* the excessive accumulation of dead keratinocytes blocking the follicle; (2) excessive secretion of sebum in response to androgens; and (3) a bacterium, *Propionibacterium acnes.* Sebum and bacteria accumulate in the blocked follicle until the follicle ruptures, exposing the contents to the dermis and triggering inflammation. The bacteria break down the lipids of sebum into free fatty acids that, combined with enzymes and other bacterial secretions, intensify the inflammation.

The primary lesion of acne is a **comedo,** a mass of keratinocytes, sebum, and bacteria in a dilated follicle. A *whitehead* is a "closed comedo" filled with sebaceous secretion; a *blackhead* is an "open comedo" that derives its darker color from oxidation of the sebum. When the infected follicle ruptures and inflammation follows, the comedo develops into a pustule, papule, or cystic nodule. The last two of these are deeper than the first and may leave a scar.

Acne is no longer thought to be caused by chocolate, sugar, or other foods. There is a hereditary influence on susceptibility to acne and its severity. Cleanliness of the skin helps minimize the severity. Acne can be treated with topical antibiotics and retinoic acid and, in severe cases, oral antibiotics. The prescription drug Acutane reduces sebaceous secretion and can markedly improve acne, but if used in the first month of pregnancy, it greatly increases the risk of birth defects. Most cases of acne clear up in a person's 20s, but the age at remission varies greatly.

Case Study 7 The Itchy Physical Therapist

Norma is a 32-year-old physical therapist working in a major hospital, where she evaluates patients and assists with their therapy. Norma works with a wide variety of patients whose disorders include skin rashes, wounds requiring debridement, and various infectious diseases that call for isolation and the use of protective clothing.

Recently Norma's hands have become red and slightly **edematous** (swollen). She has also noticed that the elastic in some garments is irritating her skin. At first, Norma attributes the irritation to the frequent hand-washing and clothing changes necessitated by her increased patient load. However, over the next few weeks, Norma notices that her hands are not improving. In fact, she is developing bullae and vesicles on her hands and at places where elastic touches her skin. Also, she is experiencing marked pruritis.

Norma makes an appointment to visit her physician the following Monday. Over the weekend, the itching decreases and the rash dissipates. After taking a history, the physician tells Norma that he thinks she may be coming into contact with something she is allergic to. He recommends skin scrapings and a patch test for some common allergies. Norma agrees, and a patch test is scheduled for the next day.

Results of the patch test indicate that Norma has developed an allergy to latex. Her physician advises her to avoid touching latex, and if she does accidentally come into contact with it or if a rash develops, to apply gauze dipped in water to the lesions 4 to 6 times a day for 30 minutes each time. He tells Norma that if blisters form, she can drain them, but she must not remove the tops of the blisters. If blisters are not present, she may use a topical corticosteroid. Finally, the physician tells her she may take antihistamines to relieve the irritation, and he mentions that latex allergies sometimes cross-react with proteins in various fruits, so she should be aware of the possibility of experiencing new food allergies.

Based on this case study and other information in this chapter, answer the following questions.

1. What risk factors are present in Norma's case? What symptoms does she have? What signs does she exhibit?
2. If Norma's doctor suspects contact dermatitis, why does he take skin scrapings?
3. Given Norma's occupation, why is it especially important that she not remove the tops of the blisters?
4. A few weeks after her diagnosis, Norma attends a party where she eats some avocado dip. Her lips tingle slightly after eating it, but she dismisses this. A month later, eating avocados at home, her lips tingle more intensely and her tongue becomes somewhat swollen. What relationship might this have to her latex allergy? What should Norma do about it?
5. How would the patch test distinguish between a latex allergy and other possible allergies?
6. Athletes who use anabolic steroids often experience increased acne. Explain why.
7. Which of the following lesions would most likely result from tinea?
   a. a wheal
   b. a nodule
   c. a fissure
   d. an ulcer
   e. scales
8. Why is the control of pruritis so important in curing skin diseases?
9. Which of the following people would be most likely to develop a pressure ulcer?
   a. a black child with the flu
   b. an elderly white man with a broken hip
   c. an Asian teenager with acne
   d. a white toddler in potty training
   e. an elderly black woman who receives dialysis treatments twice a week
10. Which of the following lesions would you most expect to see in a child with scabies?
    a. a macule
    b. a papule
    c. a bulla
    d. lichenification
    e. keloids

Selected Clinical Terms

edematous Swollen; edema is caused by the accumulation of fluid in cells or intercellular tissues.

hives Itching wheals on the skin resulting from an allergy.

infection The presence of internal pathogens.

infestation The presence of parasites or other pathogens in or on the skin.

lipoma A benign tumor of the adipose tissue.

petechia (peh-TEE-kee-uh) A small hemorrhage in the skin, pinpoint- to pinhead-sized.

seborrheic keratosis (seb-oh-REE-ik CARE-ah-TOE-sis) A greasy lesion consisting of built-up epidermal cells; often pigmented.

vitiligo (vit-ih-LYE-go) The appearance of patches of white, depigmented skin resulting from an autoimmune destruction of melanocytes in the affected area.

8 Bone Tissue

Objectives

In this chapter we will study

- methods used to diagnose bone disorders;
- three noncancerous bone diseases—osteoporosis, osteomyelitis, and osteochondrosis; and
- four forms of bone cancer—osteosarcoma, chondrosarcoma, fibrosarcoma, and myeloma.

Diagnosing Bone Disorders

As living organs, bones are subject not only to fractures but also to several other diseases. The signs and symptoms of a disorder of the bone tissue per se can be difficult to distinguish from those of other skeletomuscular diseases, thus testing the diagnostic skill of a physician. Some of the procedures described in this chapter of the clinical manual also apply to chapters 9 and 10, which deal with the skeleton as a whole and the joints, respectively.

Pain is the most common symptom of a bone disorder, but patients often wait until other symptoms arise before seeking help. Such delay may allow the disease to reach a more advanced stage or to involve additional tissues or body structures. Pain alone does little to help a clinician make the proper diagnosis. The physical examination must include careful observation of the patient's gait and posture for clues as to which bones or joints are involved. Bone abnormalities can often be detected by palpation. In addition, various tests can aid diagnosis, including the following:

- **Imaging techniques** such as X rays and CT and MRI scans may reveal dislocations, tumors, and changes in bone size.
- **Bone scans** (densitometry) help detect bone cancers, infections, necrosis, trauma, degenerative bone diseases, and metabolic disorders that affect the skeleton. To perform a bone scan, the patient is injected with a radioisotope that has an affinity for bone. After allowing the radioisotope to accumulate in the bones, the pattern of gamma-ray emission is monitored. Diseased bones show a different pattern of emission than healthy bones.
- **Angiography** is the process of making an X ray of the blood vessels. The circulatory system must be injected with a contrast medium, such as barium or iodine compounds, to enhance the visibility of the vessels on an X ray. Although this procedure is not limited to examination of the skeletal system, it is useful for evaluating the blood flow to the bones.
- **Hematologic tests** can provide clues to bone disorders by measuring the concentrations of enzymes and other chemicals in the blood serum. For example, an elevated concentration of alkaline phosphatase may indicate bone cancer or osteitis deformans. Bone cancer, fractures, and long-term immobility can raise the serum calcium concentration. Bone tumors raise the serum phosphate level.
- **Biopsy** is used to evaluate both the gross and microscopic anatomy of a small sample of bone.

Common Bone Disorders

A wide variety of disorders affect the bones, skeleton, and joints. Although these disorders can be difficult to separate, this chapter covers some of the common ones that primarily affect bone tissue itself.

Osteoporosis

Osteoporosis is the most common disorder of the skeletal system. In the United States alone, osteoporosis is estimated to cost billions of dollars per year, including treatment expenses and indirect costs such as lost work. Here we explore further aspects of this disease.

Osteoporosis is a metabolic disease in which the density of bone is reduced, but the remaining bone has a normal proportion of minerals to organic content. The histology of the bone remains normal, but the amount of bone becomes inadequate to provide mechanical support for the body. Thus, a person with osteoporosis becomes increasingly susceptible to fractures, especially of the wrist and hip. About half the people who suffer hip fractures as a result of osteoporosis never walk again.

Osteoporosis mainly involves a loss of spongy bone; compact bone is relatively unaffected. Osteoporosis is caused by aging in 95% or more of patients, but it can also occur due to hormonal imbalances, immobilization (as when a limb is in a cast), bone tumors, lack of weight-bearing exercise, and prolonged space flight. When osteoporosis affects only a particular part of the body, such as one limb, it is called **regional osteoporosis.**

Estrogen helps preserve bone mass by inhibiting the bone-dissolving action of osteoclasts, but after menopause the ovaries are inactive and estrogen is no longer secreted. In the absence of this inhibitory stimulus, bone resorption by osteoclasts increases and exceeds the bone deposition by osteoblasts. Thus, there is a net loss of bone mass.

The clinical manifestations of osteoporosis depend on the bones involved. The first symptom is pain in the bones, often described as an aching back. This pain is short-lived, but is aggravated by weight-bearing, even standing. The patient often does not seek medical attention because the pain usually subsides within a few weeks. Blood tests show relatively normal circulating concentrations of calcium and phosphorus, but parathyroid hormone (PTH) concentrations are lower than normal. Chemical indicators of bone turnover are high, including urine levels of calcium and certain collagen derivatives released by the degenerating bone. The loss of bone density can be detected by X ray, but only after 25% to 30% of the bone mass has already been lost.

Treatment of osteoporosis is aimed at preventing further bone loss and halting the progression of the disease. Dietary intakes of calcium and vitamin D are prescribed to increase the absorption of calcium. Patients are advised to limit their intake of caffeine, alcohol, nicotine, and carbonated beverages. Regular, moderate, weight-bearing exercise is recommended to slow the loss of bone and in some cases to stimulate bone formation. Hormone replacement (estrogen and progestin) is recommended to decrease osteoclast activity in postmenopausal women. Estrogen does not significantly restore lost bone, but it slows the progression of the disease by inhibiting bone resorption. However, estrogen may increase the incidence of breast and uterine cancer for some women and is therefore not always an option. Other treatments include calcitonin, given by injection or in a nasal spray, as well as oral calcitriol and other medications to minimize calcium loss. Paradoxically, intermittent low doses of parathyroid hormone can increase bone mass. This is being explored as a possible treatment for osteoporosis.

Although patients with osteoporosis should exercise, they must also take precautions to avoid falling, which can result in fractures. As an early preventive measure, weight-bearing exercise and proper nutrition are now being encouraged for young girls in the hope that depositing a greater bone mass will protect them from osteoporosis in later years.

Osteomyelitis

Bone may become infected by viruses, bacteria, fungi, and parasites; bacterial infection of bone is called **osteomyelitis.** This is an especially difficult and expensive disease to treat because when bacteria invade the bone, they are relatively sheltered from the body's disease-fighting white blood cells and antibodies. Furthermore, the bacteria may secrete toxins that promote bone necrosis, and osteocytes are unable to significantly replace necrotic bone.

Osteomyelitis is classified according to the route of bacterial invasion. *Exogenous osteomyelitis* occurs when bacteria invade from outside the body—for example, through open fractures, wounds, or surgical procedures such as joint replacement. *Endogenous osteomyelitis* occurs when bacteria, most often *Staphylococcus aureus,* invade the bone from infected sites elsewhere in the body—especially from ear, sinus, cutaneous, and dental infections. This is a common complication of sickle-cell disease and animal bites.

In children, the inflammation induced by either type of osteomyelitis causes the periosteum to move away from underlying tissues. This "lifting" of the periosteum results in decreased blood flow and the subsequent necrosis and death of the infected area of the bone. Osteoblasts surround the infected bone with new bone, but the openings in the newly synthesized bone allow pus to escape into the surrounding soft tissues. In adults, the periosteum is firmly attached to the bone surface, so this does not occur. Instead, the infection weakens the bone and makes it more susceptible to fracture.

Signs and symptoms of osteomyelitis vary depending on the infectious agent (type and source), duration (acute, subacute, chronic), site of infection, and age of the patient. Osteomyelitis is marked at first by relatively vague signs: low-grade fever, inflamed lymph nodes *(lymphadenitis),* and local pain and swelling. If the infection progresses, it causes high fever, nausea, pain, inflammation of the neighboring tissues, and muscle spasms. In chronic osteomyelitis, the long bones may develop large lesions, up to 4 cm in diameter, at their ends.

Osteomyelitis is diagnosed by means of radioisotopic bone scanning, CT, and MRI. Blood testing can also aid diagnosis since the disease typically produces an elevated leukocyte count. Osteomyelitis is treated by means of drainage and antibiotics. It may require drilling holes into the bone to allow for drainage and for antibiotics to reach the site of infection. In some instances, surgery is required to remove the exudate. If the infection occurs at the site of an artificial joint (prosthesis), the prosthesis may need to be removed in order to treat the surrounding bone.

Osteochondrosis

Osteochondrosis is a family of *avascular bone diseases* occurring in children—that is, skeletal deformities resulting from disturbances in the blood supply to the ossification centers of growing bones. It is still uncertain why the ossification centers sometimes lack a normal blood supply. Tissues around the area of bone necrosis become inflamed, joints are weakened, and bones may fracture at cartilaginous regions in and near the joints. The synovial membranes become inflamed and trigger pain and muscle spasms, often the first clinical signs of the disease. In the late stages of the disease, new blood vessels grow into the affected area and the bone is repaired. However, this restorative growth is structurally abnormal and may cause discomfort, a limp, and altered joint function. Younger children are more likely than older ones to realize a complete restoration of normal joint structure and function.

Bone Tumors

Bones are comprised of multiple tissue types, each of which can give rise to a tumor. Bone tumors are classified as osteogenic, chondrogenic, collagenic, or myelogenic, depending on whether they involve overgrowth of osseous tissue, cartilage, collagenous tissue, or bone marrow, respectively. Tumors may be *benign* or *malignant* (cancerous). All of the tumors we consider here are malignant neoplasms.

Osteogenic (bone-forming) tumors exhibit excessive growth of osseous tissue. The most common malignant osteogenic tumor is **osteosarcoma,** which accounts for 38% of all bone tumors. Osteosarcomas occur most frequently in adolescents and young adults, and they most commonly affect the humerus, femur, or tibia; half of the cases involve the knee. The tumor is usually found in the bone marrow, but it also has highly destructive effects on the surrounding bone. The growing tumor eventually breaks through the bone surface and lifts the periosteum from the bone. This triggers bizarre abnormal bone growth at the surface. The area becomes progressively painful and swollen. The tumor is treated primarily with surgery; chemotherapy and reconstructive techniques are helpful in reducing the need for amputation.

Chondrogenic tumors produce excessive growths of cartilage or a cartilage-like substance called chondroid tissue. **Chondrosarcoma,** the most common form (20% of bone tumors), usually occurs in people 50 to 70 years of age. It most often affects the femur and pelvis. The tumor is composed of large masses of hyaline cartilage and fibrous tissue. It erodes the bone, enlarges it, and often invades the joint cavity. The tumor can be surgically excised, but often returns. Therefore, amputation is often the treatment of choice.

Fibrosarcoma is a solitary collagenous tumor seen most often in the metaphysis of the femur or tibia. Its progression is from the inside of the bone out—that is, it begins in the marrow cavity, spreads to the compact bone, and eventually breaks through into the soft tissue around the bone. Fibrosarcoma often metastasizes to the lung. Radiotherapy is usually ineffective against fibrosarcoma; amputation or other radical surgery is generally necessary to save the patient.

Myeloma, responsible for about 27% of bone tumors, is the malignant proliferation of certain immune cells called *plasma cells*. About one out of six patients exhibit *multiple myeloma,* the presence of more than one tumor. Myeloma is common in people over 40 and is more common in males than in females and in blacks than in whites. The progression of myeloma is opposite that of fibrosarcoma—that is, it invades the bone from the outside in, eventually invading the marrow. It causes increasingly severe pain that is often mistaken at first for arthritis or a herniated intervertebral disc. The prognosis is poor, and patients are generally treated only to relieve discomfort. Radiotherapy and chemotherapy are not very effective against myeloma.

Case Study 8 The Bike Rider with a Broken Hip

Susan is a 55-year-old woman who regularly exercises by riding her bicycle around her neighborhood and on trails in a nearby park. One day while she is riding, a dog runs across her path, causing her to fall. Susan lands heavily on her right side and feels a sharp pain in her hip as she tries to move. When she attempts to stand, she cannot put any weight on her right leg. Some joggers help her to a nearby bench and call for assistance.

When the emergency medical technicians arrive, they immobilize Susan's leg and check for additional injuries. Upon initial examination, it appears that Susan has some abrasions on her right side and a broken hip. The emergency team transports her to the hospital. There, the emergency room physician notes that Susan's vital signs are stable and that, with the exception of her injuries, she is in relatively good health.

Suspecting a fracture of the femur, the physician orders a series of X rays to determine the site and extent of the injury. Results show a nondisplaced oblique fracture of the neck of the femur. In reviewing her medical history, the physician notes that Susan is postmenopausal and not using steroid hormone (estrogen and progestin) replacement therapy. But because she is physically active, the physician believes the fracture may be pathological (caused by a disease) and suggests further testing. When Susan asks why he suspects a problem, he tells her that, considering her age, the fracture could have resulted from either osteoporosis or a bone tumor. The only way to confirm either diagnosis is to conduct further tests. Susan consents to both a CT scan and bone densitometry. These scans reveal no tumors but a loss of bone mass, specifically spongy bone. The physician advises her to increase her dietary calcium intake to 1,500 mg/day, take vitamin D supplements, and decrease her intake of soft drinks, alcohol, caffeine, and nicotine. In addition, the physician recommends steroid replacement therapy to slow bone loss and other drugs to facilitate calcium absorption. He also encourages Susan to continue exercising once she recovers from the fracture.

The following day, surgery is performed to insert screws to internally fix the position of the femur and allow proper healing. The surgery is successful, and after a period of observation, Susan is released from the hospital. But approximately one week after the surgery, she begins to experience an increasing amount of pain in her hip. The surgical area feels warm to the touch and is redder than the surrounding tissue. Because she has an appointment with the surgeon the following day, Susan does not think she needs to visit the emergency room. But the next day at the surgeon's office, it is noted that her temperature is elevated and her lymph nodes are swollen. Blood drawn for evaluation shows an elevated concentration of lymphocytes. The surgeon orders a CT scan, which reveals a subperiosteal abscess at the surgical site.

The surgeon immediately hospitalizes Susan and prescribes antibiotics to fight the infection. In addition, a tube is inserted in the abscess to allow drainage. The surgeon tells Susan that if her condition worsens, a second surgery may be necessary to replace the screws. However, he feels that at this time antibiotics and drainage will suffice.

Based on this case study and other information in this chapter, answer the following questions.

1. Why does the emergency room physician suspect osteoporosis or bone cancer?
2. What would the CT and densitometry scans reveal if Susan had bone cancer rather than osteoporosis?
3. What is the relevance of vitamin D to Susan's prescribed course of treatment?
4. Describe the type of fracture Susan presents with.
5. Name the bone disease that develops a week after Susan's surgery. What are the signs and symptoms on which you base your answer?
6. Based on the information given, what is the likely source of Susan's bone infection?

7. John has chondrosarcoma of the tibia, and Marvin has chondrosarcoma of the pelvis. Why would John's prognosis be better than Marvin's?
8. Why would a bone tumor be a risk factor for a pathological fracture? How does a pathological fracture differ from a stress fracture?
9. Imaging techniques such as X ray, CT, and MRI are used to diagnose many bone disorders. Why do these tools play such an important role?
10. Why are osteochondroses more prevalent in children than in adults?

Selected Clinical Terms

chondrosarcoma A malignant neoplasm of cartilage or chondroid tissue of the skeleton, invading and weakening the bone.

fibrosarcoma A malignant neoplasm that arises from deep fibrous (collagenous) tissue and invades the bones and other adjacent tissues.

myeloma Malignant proliferation of plasma cells with destructive invasion of the bone.

osteochondrosis Bone necrosis, deformity, and weakness resulting from ischemia of the ossification centers.

osteomyelitis Bacterial infection of a bone.

osteoporosis A degenerative bone disease in which spongy bone is lost although the remaining bone is histologically normal; the bones become incapable of providing normal mechanical support and increasingly susceptible to fractures.

osteosarcoma A malignant neoplasm of the osseous tissue.

9 The Skeletal System

Objectives

In this chapter we will study

- two developmental disorders of the skull—acrania and craniosynostosis;
- spina bifida, a developmental disorder of the vertebral column;
- some of the causes of lower back pain later in life; and
- skeletal disorders of the feet, specifically foot deformities and heel pain.

Diagnosing Skeletal Disorders

This chapter focuses on disorders of some of the groups of bones that compose the skeleton, and examines skeletal problems at the organ and system level as opposed to the tissue level treated in the previous chapter. The methods used to diagnose the disorders discussed in this chapter are largely the same as those described in the previous chapter.

Developmental Disorders of the Skull

Developmental defects that occur during the formation of the bones of the skull can be so minor that they have little or no effect or so major that they cause death. This discussion focuses on two defects of the skull: acrania and craniosynostosis.

Acrania

Acrania ("without skull") is the almost complete absence of a calvaria, or skull cap, at birth. It is sometimes accompanied by defects in the vertebral column and by *anencephaly,* the absence or rudimentary development of the cerebrum, cerebellum, and brainstem. Acrania results from the failure of an embryonic structure called the *neural tube* to complete its development. It occurs in approximately 0.1% of live births and results in death shortly after birth.

Craniosynostosis

Normal human babies are born with unfused cranial bones that are able to shift enough to allow their heads to fit through the birth canal. The cranial bones become rigidly joined by sutures 1 to 2 years after birth. **Craniosynostosis** occurs when one or more of the cranial sutures fuses prematurely during the first 18 to 20 months of life. It occurs in about 5 out of 100,000 live births and twice as often in males as in females.

Premature closure of a suture results in lack of bone growth at right angles to the suture and compensatory growth at the sutures that remain open. For example, if the coronal suture closes prematurely, the head cannot grow normally in a fronto-occipital direction but shows excessive growth in a lateral direction, perpendicular to the sagittal suture. As a result, the head is abnormally wide (left to right) and short (front to rear). Craniosynostosis produces a deformed, sometimes asymmetric skull, often accompanied by mental retardation or other neurological dysfunctions. However, surgical intervention can limit brain damage and improve the child's appearance.

The cause of craniosynostosis is not known with certainty. One theory is that the mesenchyme—the embryonic connective tissue that gives rise to bone—lacks enzymes that normally inhibit premature ossification. This hypothesis is supported by the finding that craniosynostosis is often accompanied by other metabolic disorders.

Some abnormalities of head shape that may result from craniosynostosis are listed here, from most to least common:

- **scaphocephaly** (*scapho* = wedge), in which premature closure of the sagittal suture restricts lateral growth of the skull, resulting in a skull that is elongated vertically and in the fronto-occipital direction and has a wedgelike crown;
- **brachycephaly** (*brachy* = short), in which premature closure of the coronal suture causes excessive lateral growth of the skull, so the crown of the head is abnormally wide and the head is abnormally short from the frontal to occipital region.

- **plagiocephaly** (*plagio* = oblique), in which unilateral closure of the coronal suture causes the head to widen asymmetrically toward the side with the unclosed suture; and
- **oxycephaly** (*oxy* = sharp, pointed), in which premature closure of the coronal and sagittal sutures causes the head to be abnormally tall and narrow.

Disorders of the Vertebral Column

Disorders of the vertebral column may be congenital (resulting from abnormal fetal development and present at birth), or may develop later in life, when they are often marked by the onset of low back pain.

Spina Bifida

Spina bifida, a congenital defect of vertebral development, literally means "forked spine" (*bifid* = forked, branched). It results from a failure of the neural tube to close, especially in the lumbar to sacral region. The neural arches of the vertebrae remain incomplete, and the spinal cord is therefore incompletely enclosed in the vertebral canal. In a form called *spina bifida occulta* (*occult* = hidden), the signs are minimal and there may be little or no dysfunction. In a more serious form, *spina bifida cystica,* an infant may have a protruding sac in the lumbar region that contains spinal cord tissue. This form is accompanied by more serious neurological dysfunction.

Vertebral Disorders Causing Low Back Pain

Low back pain is pain between the inferior margin of the rib cage and the gluteal region; it may "radiate" (spread) into the thighs. Low back pain affects 60% to 80% of people at some time in their lives. The root cause is often a disorder of the vertebral column that creates pressure on the spinal nerves. Factors leading to low back pain include occupations that put stress on the vertebral column, degenerative diseases such as osteoporosis and osteomalacia, and herniated intervertebral discs, as well as the following conditions:

- **Spinal stenosis** is a narrowing of the vertebral canal or intervertebral foramina caused by hypertrophy of the vertebral bone. This condition can result from other bone disorders, such as Paget disease or osteoarthritis (see chapter 10 of this manual), and it occurs most frequently in middle-aged and elderly people. As the bone grows, it compresses the roots of the spinal nerves and may cause low back pain and muscle weakness.
- **Degenerative disc disease** occurs when the gelatinous nucleus pulposus becomes replaced by fibrocartilage with age, sometimes destabilizing the spine and leading to misalignment (subluxation) of vertebrae and ruptured discs.
- **Spondylolysis** is a condition in which the laminae of the lumbar vertebrae are defective. Defective vertebrae may be displaced anteriorly, and stresses on the vertebrae can cause microfractures in a defective lamina and eventual dissolution of the lamina.
- **Spondylolisthesis** occurs when a defective vertebra slips anteriorly, especially at the L5–S1 level. The less severe grades of spondylolisthesis may call only for palliative treatment (pain relief), but the more severe grades may require surgery to relieve pressure on the spinal nerves or to stabilize the spine.

These disorders are diagnosed through a combination of patient history, physical examination, imaging methods such as CT and MRI, tests of neuromuscular function, and other means. Most patients with acute low back pain receive short-term treatment with analgesics, rest, and physical therapy. Chronic low back pain may call for anti-inflammatory drugs, muscle relaxants, aerobic exercise, weight loss, and sometimes surgery.

Disorders of the Feet

The feet obviously bear more weight than any other part of the skeletal system. Yet each foot is composed of 26 bones. Like a machine with numerous moving parts, this creates a considerable potential for dysfunction. In this section, we discuss two types of disorders involving the feet: developmental deformities present in infancy and heel pain occurring later in life.

Foot Deformities

Foot deformities in infants may result from abnormal development or from the position of the fetus in the uterus. Early in its development, the foot normally goes through a stage of *flexion* (elevation of the toes toward the leg) and *eversion* (turning of the sole to face somewhat laterally); it usually assumes a normal position by 7 months' gestation. Developmental arrest during the early stage can lead to soft tissue deformities, including abnormally short muscles on the posterior and medial aspects of the leg, joint capsule deformities, and soft tissue contractures. The longer the condition persists, the harder it is to correct and the more likely it is to lead to rigid deformities of the bones and joints.

The two most common developmental deformities are *metatarsus adductus* and *talipes equinovarus* (clubfoot). **Metatarsus adductus** occurs in about 2 out of 1,000 live births, commonly affects both feet, and apparently results from the fetus being in a cramped intrauterine position. The infant is born with somewhat kidney-shaped feet—that is, the feet curve toward each other distally although the proximal part is normal. **Talipes equinovarus (clubfoot)** is seen in 1 or 2 out of 1,000 live births and affects both feet in about half the cases. The feet curve toward each other distally, while the heels are inverted (the soles are tilted toward each other).

Metatarsus adductus sometimes corrects itself without treatment—that is, children "grow out of it"—and in many cases it can be treated with stretching exercises and orthopedic (corrective) shoes. (Note that *orthopedic* literally means "pertaining to straight feet.") Severe cases may require the feet to be set in a series of casts that force their growth, in stages, to a correct position. Treatment of talipes equinovarus begins with foot manipulation, followed by 4 to 6 months of weekly casting, followed by the use of orthopedic shoes for 6 to 12 months. Some cases require surgery at 6 to 9 months of age to free the tight ligaments of the foot. In patients with only one clubfoot, the corrected foot often remains permanently smaller than the unaffected one.

Heel Pain

Heel pain can result from a number of disorders. Diagnosis is based on the location and type of pain as well as the age of the patient. Pain in the center of the heel may result from biochemical abnormalities, problems of the fascia (sheets of fibrous connective tissue between and overlying the muscles), or *calcaneal (heel) spurs*. In children, pain in the margins of the foot may indicate *Sever disease*. Pain posterior to the calcaneal tendon suggests Haglund deformity (enlarged posterior-superior aspect of the calcaneus). This discussion focuses on calcaneal spur syndrome and Sever disease.

Calcaneal spurs occur when a bony *exostosis* (an outgrowth of the bone surface) originates at the inner weight-bearing tuberosity of the calcaneus and extends forward toward the plantar fascia, a fibrous membrane on the plantar surface (sole) of the foot. Calcaneal spurs are thought to result from excessive pulling of the plantar fascia on the periosteum of the calcaneus. The development of a spur can cause severe pain even before the spur is detectable on X rays, but as the spur grows, the pain decreases, and a person with a fully developed calcaneal spur often feels no pain for a time. However, the pain may recur spontaneously or after trauma to the heel. At this time, diagnosis is confirmed by the appearance of a spur on an X ray. Treatment employs anti-inflammatory agents and local anesthetics. Additional relief is provided by orthopedic devices in the shoes to minimize foot elongation.

Sever disease (epiphysitis of the calcaneus) is inflammation of the epiphysis that occurs during childhood, especially in children who are very active athletically. The condition is related to the fact that the calcaneus develops from two distinct ossification centers. Ossification of the first center begins at birth, while the second ossification center appears around 8 years of age. Until complete ossification occurs (usually around 16 years of age), these centers are connected by a cartilaginous epiphyseal plate. Active children sometimes break the cartilage or the tendon that inserts on the epiphyses.

In diagnosing Sever disease, a clinician considers the patient's age, his or her level of activity, and the location of the pain. Heat and swelling in the area of the epiphysis may be signs of Sever disease. The condition is treated with heel pads in the shoes to reduce the amount of tension placed on the heel by the calcaneal tendon. Analgesics can relieve the pain, but their use is restricted in order to minimize further damage that can result from not resting the foot. In severe cases, the foot may be placed in a cast. The disease usually resolves itself over the course of several months.

Case Study 9 The Boy Whose Feet Hurt

Jamie is an active 14-year-old who rides his bike to school most days, hikes and camps with his Boy Scout troop, and plays soccer, basketball, and baseball.

One Saturday after hiking with the Scouts, Jamie complains to his parents of "sore feet." Both feet appear somewhat swollen and are slightly warm to the touch. His parents ask if he hurt his feet on the hike. Jamie remembers that he slipped during the hike and landed awkwardly, but nothing fell on his feet. Jamie's parents tell him to rest for the remainder of the day and he will probably be fine the next morning. On Sunday, the swelling has decreased, and Jamie resumes his normal activities.

A week later, Jamie again complains that his feet are sore and that his favorite hiking boots feel too tight. Growing concerned, his parents tell him to rest and to refrain from athletic activities until he can be checked out. They make an appointment with his pediatrician for the next week.

During the examination, the pediatrician notes Jamie's activities and relative absence of injuries. Jamie's vital signs are all within normal range for a 14-year-old boy, and his feet do not appear swollen or inflamed. X rays of his feet show no broken bones. Based on these results, the pediatrician suggests that Jamie has unduly stressed his feet by participating in so many activities. He recommends a week of rest and tells Jamie's father to schedule another appointment if the pain returns.

After resting for the prescribed length of time, Jamie again resumes his normal activities. Unfortunately, the pain returns, so he goes to see his pediatrician the next day. Upon examination, the doctor notes that Jamie's feet are swollen, warm, and tender to the touch. In addition, the pain seems to be localized along the heel. X rays for fractures are again negative, and no bone spurs are apparent. Tests for other childhood diseases and juvenile rheumatoid arthritis are also negative.

Based on these signs and the absence of other diseases, the pediatrician diagnoses Jamie with Sever disease. He explains that this condition is common in children between the ages of 8 and 16, especially those who are very active. Treatment will consist of placing special pads in the heels of Jamie's shoes to relieve the strain on the tendon in his heel and limited use of aspirin to alleviate the pain. The pediatrician tells Jamie that the pain should subside within a few months and that this condition is a normal part of growing up for some children. However, he cautions Jamie to be somewhat careful and to come back if the pain persists. Jamie's father asks the pediatrician if Jamie can take additional aspirin as needed for the pain. The doctor answers, "No—we want him to feel *some* pain."

Based on this case study and other information in this chapter, answer the following questions.

1. Why isn't Jamie allowed unlimited use of aspirin for pain?
2. Why does Jamie's pain subside after a period of inactivity?
3. Why is Sever disease more likely to occur in a child who is active than in one who is sedentary?
4. Why does Sever disease occur in children rather than adults?
5. Describe the pattern of pain appearance and remission that can be expected if Jamie uses heel pads only intermittently.
6. Why are Jamie's X rays normal? If Jamie had calcaneal spurs, how would the X rays have differed?
7. In craniosynostosis, why would premature closure of the sagittal suture restrict the lateral growth of the skull?
8. If you were a pediatrician, what measurements of the skull would you take to distinguish different forms of craniosynostosis from each other?
9. About 50% of children with craniosynostosis exhibit mental retardation. Explain the probable connection between the skeletal deformity and the neurological effect.
10. Explain why Paget disease could cause abnormal pressure on a spinal nerve.

Selected Clinical Terms

acrania The congenital absence of most of the calvaria of the skull, often accompanied by defects in spinal and brain development.

calcaneal spur A bony outgrowth of the calcaneus that can cause severe foot pain.

craniosynostosis The premature fusion of one or more cranial sutures, resulting in disrupted cranial growth and various skull deformities.

metatarsus adductus A congenital foot deformity in which the distal part of the foot curves medially and the proximal part is normal, giving the foot a kidney shape.

Sever disease Inflammation of the epiphyseal plate in the calcaneus of children.

spina bifida Incomplete development of the neural arches of the vertebrae, sometimes resulting in protrusion of spinal cord tissue from the vertebral canal and causing severe neurological dysfunction.

talipes equinovarus Clubfoot; a rigid, congenital foot deformity in which the distal part of the foot curves medially and the heels are inverted.

10 Joints

Objectives

In this chapter we will study

- clinical signs and diagnostic tests for joint disorders in general; and
- several specific joint disorders—osteoarthritis, rheumatoid arthritis, ankylosing spondylitis,
- bursitis, and hip dysplasia.

Diagnosing Joint Disorders

The bones of the skeleton are connected by joints of several types, each having a specific form and function.

The diagnosis of joint disorders is based on many of the same signs, symptoms, and diagnostic methods as the disorders of the skeletal system described in the two previous chapters. In addition, however, the following criteria are relevant to joint disorders in particular:

- **Range of motion (ROM)** is a measurement of the degrees through which a joint can be moved; in effect, it quantifies the flexibility versus the stiffness of a joint. ROM is measured with a device called a goniometer. The *active range of motion* is the range through which a patient can move a joint by his or her own effort; the *passive range of motion* is the range through which an examiner can move the patient's joint. The ROM observed in a physical examination can be compared with the population norm or with previous measurements on the same patient to monitor the severity and course of disease or the progress of therapy.
- **Crepitus** is a crackling or grating sound or a vibration produced by joint movement.
- **Arthrography** is a diagnostic procedure in which dye is injected into a joint and then an X ray is taken. This method can help identify such anatomical disorders as tears in a meniscus of the knee or damage to the rotator cuff of the shoulder.
- **Arthroscopy** is the viewing of the interior of a joint with an arthroscope. This is a common way of looking for damaged cartilage and ligaments in the knee and other joints and for viewing the joint cavity during surgery.
- **Synovial fluid sampling** is conducted by aspirating a specimen of fluid from a joint and examining it for chemical changes or the presence of bacteria. This test can help to distinguish septic, inflammatory, and noninflammatory joint diseases. Cell fragments or fibrous tissue in the synovial fluid can indicate excessive wear on the articular surfaces. **Hemarthrosis,** blood in the synovial fluid, suggests joint trauma.

Arthritis

Arthritis is a broad term that describes painful or inflamed joints. Although many of us associate arthritis with synovial joints such as the knees and fingers, all types of joints are susceptible to arthritis. There are two principal forms of arthritis: osteoarthritis and rheumatoid arthritis.

Osteoarthritis

Osteoarthritis (OA) is the most common noninflammatory joint disease, affecting about 85% of people over 70 years of age. Although changes in the weight-bearing joints begin in some individuals as early as the second decade of life and all people experience some joint pathology by age 40, the symptoms usually don't show up until age 50 or 60. Men and women are equally affected, but men demonstrate an earlier age of onset and women suffer more severe effects. Interestingly, almost all vertebrates are susceptible to OA. Fossilized vertebrates from fish to mammals show signs of it, as do marine mammals, even though their joints bear less weight than those of terrestrial mammals. However, bats and sloths, which spend most of their lives hanging upside down, do not exhibit OA.

Osteoarthritis is subdivided into idiopathic (primary) and secondary types. *Idiopathic OA* cannot be traced to an identifiable cause, whereas *secondary*

OA is associated with known risk factors such as conditions that damage cartilage, cause joint instability, or place chronic or abnormal force on the joint surfaces. Risk factors for secondary OA include the following:

- joint trauma such as sprains, strains, and dislocations;
- repeated mechanical stress due to a physical task (including athletics and ballet);
- inflammatory joint disease, which releases enzymes that degrade articular cartilage;
- neurologic disorders that diminish the sense of pain, so that a person engages in abnormal movements or body positions that would ordinarily be painful (diabetes mellitus is one such pain-reducing pathology);
- skeletal deformities (congenital or acquired);
- drugs such as colchicine or steroids, which increase the activity of enzymes that digest collagen in the synovial fluid; and
- disorders such as hemophilia that cause chronic bleeding into a joint.

Both idiopathic and secondary OA cause a loss of articular cartilage. Initially, the cartilage changes from white and glistening to yellow-gray or brownish-gray. The cartilage becomes thinner and marked by delicate fissures *(fibrillation)*. The complete loss of articular cartilage from some areas of the joint exposes the underlying *(subchondral)* bone, which subsequently exhibits thickening and hardening *(sclerosis)*. Bone spurs *(osteophytes)* grow from the surface, alter joint anatomy, and impair joint movement.

The most common symptom of OA is joint pain, especially in the hand, wrist, neck, lower back, hip, knee, ankle, or foot. The onset of the pain is gradual. Joint stiffness is typically noticed first thing in the morning and disappears after a few minutes of movement. OA tends to be aggravated by cold and high humidity. As the disease progresses, range of motion in the joint is decreased, joint tenderness and crepitus are noted, and the patient may complain of a grating sensation in the joint. These symptoms are followed by joint enlargement, tenderness, and deformity.

Diagnosis is usually based on the patient's symptoms. X ray, CT, MRI, and arthroscopy are often used to confirm a diagnosis and may lead to the discovery of OA in asymptomatic patients. Blood studies are not used to diagnose OA, but they help rule out other possible causes of similar symptoms.

Treatment usually involves resting the affected joint if it is inflamed; range-of-motion exercises to prevent contracture of the joint capsule; weight loss if the patient is obese; using a cane, crutches, or a walker to take some of the weight off a joint; and taking analgesic and anti-inflammatory drugs to reduce the pain and swelling. In severe cases, a joint may have to be surgically replaced—especially the hip or knee. In the United States alone, 250,000 patients undergo total hip replacement each year, usually because of OA.

Rheumatoid Arthritis

Rheumatoid arthritis (RA) is an autoimmune disease that occurs when the body develops antibodies (rheumatoid factor, RF) against the synovial membranes. The inflammatory response spreads from the synovial membranes to the articular cartilage, fibrous joint capsule, and surrounding ligaments and tendons. This results in pain, joint deformity, and loss of function. RA affects especially the fingers, feet, wrists, elbows, ankles, and knees, and sometimes the shoulders, hips, and cervical spine. The etiology of RA is not completely understood, but it is believed to involve a combination of genetic, environmental, and hormonal factors. Whatever the cause, once RF is produced, it sets off a series of events leading to the signs and symptoms of RA.

RA affects 1% to 2% of adults and three times as many women as men. The most common age at onset is between 25 and 50 years, but it can strike patients of any age, including children. In fact, 40 of every 100,000 children develop juvenile RA (JRA), which accounts for 5% of all RA cases. For the most part, JRA and RA have similar clinical manifestations but different modes of onset.

The onset of RA is often abrupt and usually occurs in many joints simultaneously. However, in some cases the onset may take place gradually over a period of weeks or months. The patient complains of tenderness in the affected joints as well as feeling stiff in the morning or after prolonged inactivity (longer than 30 minutes). The patient may also experience fever, fatigue, weakness, weight loss, anorexia, and generalized aching and stiffness. The skin over the affected joints may appear thin and shiny, with a reddish-blue tint. As the pain in the joint limits the range of motion, permanent deformities of the fingers, toes, and limbs may occur. The loss of ROM is accompanied by atrophy of the surrounding muscles, which leads to increased joint instability. As the inflammation worsens, *rheumatoid nodules,* or swellings, are sometimes seen in areas such as the extensor surfaces of the elbows and fingers. Rheumatoid nodules can also occur on the scalp, back, buttocks, and even in the heart, lungs, and spleen. Each nodule consists of a fibrous capsule around a core of necrotic tissue.

Diagnosis is based on the patient's history, signs and symptoms, imaging tests, and laboratory results. X rays or other images show bone erosion and narrowing of the joint cavity. Blood tests are positive for rheumatoid factor and often reveal anemia. The synovial fluid has a cloudy appearance and contains an abnormally large number of leukocytes.

As with many diseases, there is no cure for RA, so the treatment goal is palliative. Treatments include rest, exercise, physical therapy, immunosuppressive drugs, and various anti-inflammatory drugs. Surgery may be employed to correct deformities that arise.

Ankylosing Spondylitis

Ankylosing spondylitis (AS) is a debilitating joint disease affecting primarily the sacroiliac and intervertebral joints. It is similar to rheumatoid arthritis in that both conditions are inflammatory diseases involving the immune system. However, RA affects mainly the synovial membrane, whereas AS involves the point at which ligaments, tendons, and joint capsules insert into the bone. AS leads to fusion of the joints **(ankylosis),** especially the sacroiliac and intervertebral joints. Although men and women are equally affected, AS occurs more frequently in Native Americans (18% to 50% incidence) than in blacks (3% to 4%) and whites (0.5% to 1%). However, these data may not accurately reflect the true incidence of AS because it is believed that most individuals with AS are undiagnosed.

The exact cause of AS is unknown, but a genetic factor has been noted in some individuals. *Primary AS* has an early onset (late adolescence to young adulthood), while *secondary AS,* associated with other inflammatory diseases, affects older age groups.

The earliest symptoms of AS are lower back pain, especially at night, and stiffness, especially early in the morning. The pain is intermittent at first, but progressively becomes more prolonged, while the stiffness seems to resolve with activity. Other early symptoms include fatigue, anorexia, weight loss, anemia, and fever. As the disease progresses, compression of the spinal nerves results in characteristic neurologic signs such as nocturnal urinary incontinence, decreased rectal and bladder sensations, and absence of the ankle reflex (plantar flexion of the foot in response to sudden stretching of the calcaneal tendon).

AS is typically diagnosed through patient history, physical examination, blood testing, and X rays. The blood shows an elevated concentration of alkaline phosphatase and a higher than normal erythrocyte sedimentation rate (ESR; the rate at which erythrocytes settle in a sample of blood). Although these signs are suggestive of AS, diagnosis can be confirmed by X rays. The vertebrae often lose the normal concave shape of the anterior surface of the centrum, becoming cylindrical, which leads to a characteristic "bamboo spine." Sclerosis and erosion in the sacroiliac joint are also common. Patients often acquire a "stooped" posture as the lumbar spine becomes straight and the thoracic spine becomes more rounded *(kyphosis).*

Once a diagnosis of AS is confirmed, treatment is initiated to alleviate the pain using anti-inflammatory drugs, heat, and therapeutic massage. These treatments are usually combined with physical therapy to maintain posture, range of motion, and normal chest expansion in breathing. Improving (or maintaining) posture minimizes deformation and lack of mobility. If diagnosed early, the prognosis is good. Most AS patients lead normal lives with minimal or no disability.

Bursitis

Bursitis is inflammation of a *bursa.* It can be acute or chronic. The most common cause is overuse of the joint, but infection, trauma, and inflammation can also stimulate the condition. Bursitis most often affects the shoulder, but can also involve other joints. An inflamed bursa becomes engorged with fluid and may stimulate inflammation of adjacent tissues. In chronic bursitis, the bursal wall thickens and often develops adhesions and calcareous deposits. Bursitis is characterized by pain and decreased range of motion in the affected joint. The pain tends to be localized to a certain region and is more severe when the joint is moved. For example, shoulder bursitis limits abduction of the arm, while hip bursitis creates pain when the legs are crossed.

Diagnosis of bursitis relies heavily on physical examination and the patient's history. Localized tenderness or swelling of a bursa is often the first sign. Imaging techniques may be used to confirm the diagnosis. Bursitis is most often treated by immobilizing the limb and administering anti-inflammatory drugs. In cases of chronic bursitis, surgical removal of adhesions or calcium deposits may be required. This is followed by physical therapy to return range of motion and minimize muscle atrophy.

Developmental Dysplasia of the Hip

Developmental dysplasia of the hip (DDH), formerly known as *congenital dislocation of the hip,* is an abnormal development of the proximal femur, acetabulum, or both, resulting in an unstable coxal joint and sometimes hip dislocation. It is usually evident at birth, but may appear at any time in infancy. The cause is still obscure, but it has a hereditary component and affects six times as many girls as boys. It is sometimes associated with conditions that limit fetal movement or put unusual stress on the hips, such as a deficiency of amniotic fluid or a breech presentation instead of the normal head-down fetal position.

Infants with DDH typically exhibit subluxation of the hip. **Subluxation** is a partial dislocation in which two bones maintain contact between their articular surfaces but are not entirely dislocated. In subluxation of the coxal joint, the head of the femur remains in contact with the acetabulum but is not properly inserted into it. The acetabulum is often shallow or sloping rather than cuplike. The subluxated joint is easily subject to true dislocation *(luxation),* in which the articular surfaces of the two bones lose contact. If DDH is not diagnosed early enough, it can progress to secondary changes in the bone and adjacent soft tissues, leading to a disrupted blood supply to the hip, flattening of the femoral head, stretching of the joint capsule, and deformity of the acetabulum.

The severity of the dysplasia and the age of the child influence the signs and symptoms of DDH. In general, these include pain, differences in the lengths of the legs, limited hip abduction, and a waddling gait.

The treatment regimen depends on the child's age and the severity and duration of the dysplasia. The earlier treatment is initiated, the better the prognosis. If the child is not yet walking, treatment consists of placing the hip in a flexed and abducted position and maintaining it with braces, traction, and casts, alone or in combination. The joint position must be maintained until the joint capsule tightens and the acetabulum assumes a more normal shape. Treatment of dysplasia in children who are walking usually requires surgery and casting.

Case Study 10 The Woman with Sore Knees

Angela, a 72-year-old widow, visits her physician complaining of pain in her knees. She says the pain seems to worsen if it is raining or snowing, and is most severe in her left knee, the one she twisted 4 years ago when she fell and broke her hip on an icy sidewalk. She became more aware of her sore knees when she began having difficulty climbing the stairs in her home. She also mentions that her back has been bothering her for so long that she can't remember when she didn't have pain there.

In reviewing Angela's medical history, the physician notes that she has gained 15 pounds since her last checkup, does not complain of any swollen, inflamed, or deformed joints, and is experiencing no problems with her hips. Angela mentions that while playing tennis in college she twisted her left knee, but never needed surgery. She also says that although her joints are generally stiff in the morning or after she has been sitting for awhile, they seem to "loosen up" after she begins to move around. When asked about numbness, pain, or weakness in her legs, she replies that she has not experienced any of those symptoms.

The physician continues to review Angela's record and to ask questions. She notes that Angela did not choose hormone replacement therapy following menopause, has no allergies, and is not currently taking any prescription drugs. Angela does report taking ibuprofen for her pain, but says she does not take it regularly because it upsets her stomach. Angela does not smoke, and she exercises as often as possible, although she cannot walk as far as she used to because of the pain in her knees. She does not drink alcoholic beverages, and her diet is balanced but high in processed foods and low in calcium. Results of her physical examination and laboratory tests are shown here.

Mental status: Alert

Oral temperature: 98.8°F (37.1°C)

Heart rate: 70 beats/min

Respiratory rate: 12 breaths/min

Blood pressure: 160/84 mmHg

Musculoskeletal examination: Full passive range of motion (ROM) at shoulders and elbows. Decreased ROM in both hands. Decreased flexion and extension of the back, with mild scoliosis. Enlargement of both knees with decreased ROM and crepitus, greater in the left than in the right. Left hip with decreased rotation, but pain-free.

Laboratory tests: All normal.

X rays: Left knee shows narrowing of joint space, subchondral sclerosis, and bone cysts with no apparent osteoporosis; lumbrosacral spine shows narrowing disc space and no compression fractures.

Based on these results and Angela's symptoms, her physician diagnoses osteoarthritis. She suggests that Angela attempt to lose weight gradually and increase her calcium intake. In addition, she should continue to exercise moderately and begin physical therapy to try to keep the condition from worsening. The physician also suggests that Angela switch to acetaminophen for pain to minimize stomach upset and consider taking chondroitin sulfate and glucosamine sulfate, which are *chondroprotective drugs* that help maintain cartilage and promote the repair of damaged cartilage. Finally, Angela's doctor suggests both bone scanning and hormone replacement to monitor the possible development of osteoporosis.

Based on this case study and other information in this chapter, answer the following questions.

1. What signs and symptoms support the diagnosis of OA in Angela's case? How is RA ruled out? How is osteoporosis ruled out?
2. Why does Angela's doctor recommend exercise and weight loss?
3. Why might chondroitin sulfate and glucosamine sulfate improve Angela's condition?
4. Even though Angela is not diagnosed with osteoporosis, which of the treatments recommended by her doctor are related to that condition? Why does the doctor make those recommendations?
5. Would crepitus be detected by palpation, auscultation, or percussion? Explain.
6. Why would decreased range of motion indicate a joint disorder rather than a skeletal disorder?
7. Why do you think sloths and bats are not susceptible to osteoarthritis?
8. Why does ankylosing spondylitis cause postural changes?
9. Which one of the following joints would *not* be susceptible to bursitis?
   a. humeroscapular joint
   b. intercarpal joint
   c. tibiofibular joint
   d. radiocarpal joint
   e. coxal joint
10. In older infants and children, developmental dysplasia of the hip can be confirmed by X ray. In newborns, however, this method is ineffective, so ultrasound is used instead. Explain the reason for this difference in diagnostic imaging.

Selected Clinical Terms

ankylosis The pathological fusion of two bones across a joint, resulting in stiffening or immobility of the joint.

bursitis Inflammation of a bursa.

crepitus Noise or vibration produced when bony or cartilaginous surfaces rub across each other.

dysplasia Abnormal development of any tissue or organ.

osteoarthritis (OA) A degenerative joint disease in which articular cartilage is eroded and the bone exhibits abnormal sclerosis or spurs, resulting in pain and loss of joint function.

rheumatoid arthritis (RA) A systemic autoimmune disease of the connective tissues that produces its most noticeable effects in the joints; involves an attack of autoantibodies called rheumatoid factor on the synovial membranes, leading to pain, joint deformity, and loss of function.

subluxation A partial misalignment of two bones such that their articular surfaces maintain contact but do not properly articulate with each other. May progress to true dislocation (luxation) in which the articular surfaces lose contact.

11 The Muscular System

Objectives

In this chapter we will study

- methods used to diagnose muscle disorders; and
- some structural disorders of skeletal muscle—rhabdomyolysis, myositis, and the muscle tumor rhabdomyosarcoma.

Diagnosing Structural Disorders of the Muscular System

The body contains about 600 skeletal muscles, specialized organs that produce movement of body parts and perform several other essential and overlapping functions. Health-care professionals employ a knowledge of muscle form and function when they turn patients, give injections, or engage in the diagnosis and treatment of muscular system disorders **(myopathies).** Although muscle structure and function cannot be strictly separated, we deal with myopathies of a primarily structural nature in this chapter and with more functional disorders in chapter 12.

The symptoms of muscular disorders can be confusing because some of them, such as pain upon movement, are also produced by disorders of the nervous and skeletal systems. A skilled clinician must be able to determine which system is responsible for a given symptom so that proper treatment can be initiated.

The most common symptom of a muscular disorder is pain in the affected muscle or muscles; the second most prevalent symptom is weakness. Either of these symptoms may be due to nervous or skeletal system pathologies, trauma to the muscle itself, or muscle infections. Additional symptoms include muscle atrophy, tenderness, stiffness, and cramping. When a patient presents with these rather broad symptoms, how does a clinician complete the diagnostic process?

During the physical examination, observing the patient's gait and posture can help detect disorders of muscles in the limbs or trunk, while assessing the patient's speech and facial movements provides information about the musculature of the head. In addition, various diagnostic tests can be conducted to evaluate muscle strength and motion.

Often muscle strength is first tested manually. Conducting this exam requires the clinician to understand muscle anatomy and the functional relationships of the muscles in a given region. For example, to test the quadriceps femoris muscle, which extends the knee, the clinician may ask the patient to sit on an examination table and then lift and straighten his knee. This will show whether the patient can lift his leg against gravity. If he can do this, the clinician may next push down on the lower leg as the patient tries to straighten his knee against this resistance. This exercise tests the muscle's ability to work against an externally applied force. The clinician judges the amount of force the patient can generate and assigns a grade according to the guidelines followed at the particular clinic. There are three commonly used scales: (1) a percentage scale, which rates muscle strength between 0% and 100%; (2) a terminology-based scale, in which muscle strength can range anywhere between "trace/zero" and "good/ normal"; and (3) a numerical scale, which ranges from 1 for "no contraction" to 5 for "normal contraction strength." To further differentiate muscle strength when using this scale, a plus (+) or minus (–) may be added to the number grade.

In clinical settings, muscle force is also frequently evaluated using various devices called **myometers** or **dynamometers,** which measure the extent and force of muscle contraction at individual joints. As noted in chapter 10, the goniometer measures the range of motion of a joint in degrees.

Other tests that may help diagnose muscular disorders include magnetic resonance imaging (MRI) and electromyography. An MRI can distinguish between soft tissues, making it extremely useful in muscle evaluations. **Electromyography** is the recording of the electrical activity of a muscle, using either a needle electrode inserted into the muscle or electrodes applied to the skin overlying the muscle. Electromyography can help differentiate muscle diseases from diseases of the nervous system or of the neuromuscular junction.

Muscle biopsies are sometimes performed to diagnose muscle infections and diseases that alter muscle structure. In this procedure, a small sample of muscle tissue is removed and then examined with a light microscope to detect histological or histochemical abnormalities or with an electron microscope to assess changes in cell ultrastructure (the thick and thin filaments, organelles, and plasma membrane).

A battery of blood and urine tests can also help identify muscle dysfunctions. Important measurements include:

- muscle enzymes that have leaked into the blood, such as creatine kinase (CK), lactate dehydrogenase (LDH), aldolase, or aspartate transaminase;
- changes in the concentration of electrolytes (such as calcium and potassium), which are normally more concentrated in the skeletal muscle than in the blood; and
- the concentration of muscle myoglobin in the urine.

The normal ranges for these clinical tests are given in the Appendix of Normal Values at the end of this manual. The most useful of these tests measures *serum CK* concentration. Because this enzyme is found in high quantities in muscle fibers, muscle damage releases CK into the tissue fluid and consequently into the blood. Like CK, myoglobin is released from muscle cells by trauma, ischemia, extreme exertion, and certain genetic disorders; measurable amounts quickly appear in the urine. The *serum aldolase* level can help distinguish between muscular dystrophy, which raises the aldolase level, and such diseases as myasthenia gravis and multiple sclerosis, which do not.

Changes in *serum electrolyte* concentrations can result in muscle pathologies, so measuring the electrolytes provides information about the possible cause of a change in muscle function. For example, muscle weakness can stem from decreased serum potassium or elevated serum calcium concentrations, while muscle twitches and tetany can be induced by decreased serum calcium concentrations.

Finally, the *forearm ischemic exercise test* enables a clinician to test the enzymatic pathways used for energy production in the muscle. The patient performs a series of strenuous exercises employing a maximum grip strength and graded workloads. Blood is drawn for up to 20 minutes after the conclusion of the exercise and then analyzed for lactate and serum enzymes. The exercise induces lactic acid fermentation and raises the lactate level in the blood, but an excessively elevated lactate level may indicate mitochondrial disorders, and an abnormally low lactate level suggests disorders of the glycolytic pathway. Ammonia levels normally rise along with the lactate level, but if no increase in ammonia concentration is seen, this suggests a deficiency of the enzyme monoadenylate deaminase, which is important in making creatine phosphate and ATP. A deficiency of this enzyme can make a person unable to tolerate exercise.

Structural Disorders of Skeletal Muscle

Rhabdomyolysis

Rhabdomyolysis is a severe, potentially fatal destruction of muscle cells. The breakup of muscle cells releases myoglobin, which soon appears in the urine, a condition called *myoglobinuria.* In fact, myoglobinuria is such a common sign of rhabdomyolysis that the two terms are sometimes used synonymously.

The presence of myoglobin gives the urine a dark, reddish-brown color. Damage to as little as 200 grams of muscle is sufficient to induce a change in the appearance of the urine. Serum CK is also elevated in myoglobinuria, often to a concentration 100 times greater than normal.

The most severe form of rhabdomyolysis is **crush syndrome**—massive destruction of muscle tissue resulting from major trauma such as war injuries, getting a limb caught in farm or factory machinery, or being run over by a vehicle. Such serious injuries can release enough myoglobin to block kidney tubules and cause death from renal failure. Another form of rhabdomyolysis is **compartment syndrome,** which involves selective injury to the muscles within a particular compartment of the upper or lower limb. Rhabdomyolysis can also be caused by viral infections, bacterial tetanus, heatstroke, fractures, excessive muscular activity, or certain anesthetics.

When a limb is injured, the mere weight of the immobile body part can cause muscle ischemia. The resulting decrease in venous drainage allows fluid to accumulate in the limb, causing edema (swelling). This edema causes the pressure within the compartment to rise, and if not relieved, it may further compress the blood vessels in the compartment, reduce blood flow, and cause cells to die from a lack of oxygen or from accumulation of metabolic wastes. As muscular tissue dies, it is replaced by scar tissue. This can lead to an abnormal shortening of the muscle called **contracture.**

Management of rhabdomyolysis is aimed at diagnosing and treating the underlying cause. It is particularly urgent, however, to relieve the pressure in the muscle compartment and to prevent or reverse the myoglobinuria, which can cause renal failure if left untreated. Compartment pressure can be reduced by fasciotomy (cutting the muscle fascia) to relieve pressure and, if necessary, debridement (removing dead and nonviable tissue from the injury site).

Myositis

Myositis is any form of muscle inflammation. It may be caused by a viral, bacterial, or parasitic infection. Three types of myositis are described here: trichinosis, polymyositis, and dermatomyositis.

Trichinosis is caused by the larvae of several species of parasitic roundworms in the genus *Trichinella.* These larvae can be found encysted in the muscles of infected mammals, including humans, dogs, pigs, horses, and wild game. People are most commonly exposed when they eat undercooked meat, especially pork, but also wild game such as bears and seals. Thorough cooking or freezing destroys the larvae and eliminates the risk of trichinosis. After being ingested, the larvae migrate to the intestinal mucosa where they develop, mate, and produce new larvae. These new larvae then move from the intestine into the lymphatic and circulatory systems, which transport them to muscles throughout the body. Once there, the larvae penetrate individual muscle fibers and encyst.

Symptoms of trichinosis include severe muscle and joint pain, generalized muscle stiffness, rash, and edema. Diagnosis is often accomplished through muscle biopsy. It is estimated that up to 4% of the U.S. population is infected with *Trichinella* at any given point in time. Of the people who experience symptoms severe enough to require treatment, approximately 1% die. Death usually results from tissue damage caused by larvae migrating through the heart, kidneys, and respiratory muscles.

Polymyositis is an inflammation of multiple muscles at one time; it is sometimes accompanied by characteristic skin lesions and is then called *dermatomyositis.* Although these two conditions affect only 6 out of 1 million people, they are the most frequent types of myositis requiring chronic care. The causative agent has not been identified, but it is believed to affect the immune system as well as the muscular system.

Both polymyositis and dermatomyositis are characterized by inflammation of the muscle and associated connective tissue, which is often severe enough to induce necrosis and destroy muscle fibers. Both diseases produce a wide variety of symptoms, including swollen, tender, or painful muscles, fever, lethargy, and malaise. In addition to these symptoms, symmetrical muscle weakness occurs—that is, the corresponding muscles on both sides of the body are similarly weak. Dermatomyositis is characterized by a purple-colored rash on the face, chest, eyelids, and upper limbs.

Because the symptoms are so nonspecific, both diseases are difficult to diagnose, and the clinician must complete a thorough examination in order to rule out other diseases with similar symptoms. The most helpful diagnostic procedures are muscle biopsy, MRI, and clinical laboratory tests. Definitive signs of dermatomyositis revealed by a biopsy are atrophied muscle fascicles and a large number of inflammatory cells surrounding the blood vessels. MRI results support the biopsy findings, showing inflammation and muscle edema. The primary clinical laboratory finding is an elevated serum CK concentration.

Treatments for polymyositis and dermatomyositis center on reducing the inflammation (often by administering anti-inflammatory drugs) and prescribing physical therapy to minimize muscle atrophy and contractures.

Muscle Tumors

Muscle tumors occur most frequently in children. They may arise either from muscle tissue or from associated tissues such as adipose tissue, synovial membranes, or nervous tissue.

Rhabdomyoma is a rare, benign muscle tumor. It most often occurs in the tongue, neck, larynx, nasal cavity, uvula, heart, and vulva. The treatment is simply to remove the tumor surgically; they usually do not recur.

Malignant skeletal muscle tumors, termed **rhabdomyosarcomas (RMS),** are seldom seen in adults. Even in children, RMS accounts for less than 3% of cancer cases, but it is the most common form of pediatric soft-tissue sarcoma. It usually appears at the age of 2 to 6 years or 15 to 19 years, with no difference between the sexes.

During fetal development, some embryonic myoblasts (the embryonic cells that give rise to skeletal muscle fibers) become cancerous rather than differentiating into normal striated muscle. These abnormal myoblasts are called *rhabdomyoblasts*. The cause of the transformation is as yet unknown, but one tumor suppressor gene (*p53*) and three oncogenes (c-*myb, src,* and H/K-*ras*) may be involved. Any skeletal muscle can develop a tumor, but RMS is most common in the head and neck, trunk, limbs, and urogenital tract. RMS metastasizes frequently and rapidly to the lymph nodes, bone, bone marrow, brain, liver, heart, and lungs.

The signs and symptoms of RMS are determined by tumor location. In most cases, there is no pain, and the tumor is first detected when a visible or palpable mass is noticed. If a tumor is suspected, imaging techniques are used to confirm its location and possible metastatic sites. Final diagnosis is achieved by biopsy. Metastasis and organ involvement are confirmed by blood tests and organ-specific evaluations, such as measuring liver function or obtaining bone marrow by aspiration (suction) for microscopic examination.

Treatment combines surgery, chemotherapy, and radiation therapy. If the tumor is localized (has not spread), the survival rate is high (70–80%). However, with widespread metastasis, the probability of survival drops to 20%.

Case Study 11 The Pitcher with a Sore Arm

Jason, a 20-year-old college junior, is getting ready for the start of the baseball season by working out with some friends over the winter. Since he is hoping to be selected as one of the top two starting pitchers, he spends many hours in the weight room improving his strength and conditioning. In addition, he has a friend catch for him each day. Jason starts working out slowly and carefully because he is concerned about overtraining and incurring injury before he has gotten in shape.

As the start of the baseball season approaches, Jason does not feel he has progressed enough, so he increases the duration and intensity of his workouts. He also decides that his conditioning is sufficient to spend most of his time pitching. This approach seems to work because the coaches choose him as one of the starting pitchers for the team.

About five games into the season, Jason notices that his pitching arm tires more quickly than it did earlier. At the beginning of the season, he felt he could pitch an entire game, but now he is tiring by the sixth or seventh inning. He also notices that his fastball no longer has its "zip," so he begins relying more on his curve ball. This tactic seems successful in the next two games, but by the eighth game of the season, Jason can't get any of his pitches over the plate with enough velocity to keep the batters from hitting them. The coach removes him from the game in the second inning and sends him to the team physician for evaluation.

After obtaining a complete history, the physician conducts a physical examination and orders a series of tests, including evaluation of serum enzyme and electrolyte concentrations, an MRI of Jason's pitching shoulder, and range-of-motion and strength tests on the joint. The results of these tests are presented here.

Visual examination: Right shoulder "droop" and lack of free movement of right arm during walking. Right shoulder and arm held more closely to body.

Muscle strength: All normal with the exception of the right arm rotator cuff muscles, which are given a grade of 4– in a manual test.

Range of motion: Normal for all areas tested except for the right shoulder, which is reduced and guarded.

MRI: Inflammation of muscles of the right shoulder and arm.

Laboratory tests:
- Serum creatine kinase (CK) = 200 IU/L
- Serum lactate dehydrogenase (LDH) = 190 IU/L
- Serum potassium = 4.2 mEq/L
- Serum calcium = 0.2 mEq/L

Based on the test results, the team physician diagnoses a strained rotator cuff. He tells Jason to rest his shoulder for the next 2 weeks and then come back for reevaluation. In the meantime, he recommends that Jason ice his shoulder, wrap it with an elastic bandage, and use a sling. The physician also advises Jason to take Motrin twice daily to reduce the pain and inflammation.

Upon reevaluation 2 weeks later, the physician allows Jason to return to practice but cautions him about overexertion. He warns Jason that since his rotator cuff has been inflamed, the chances of more serious damage, such as tearing, are greater for him than for other pitchers.

Based on this case study and other information in this chapter, answer the following questions.

1. Why is Jason given an MRI scan rather than an X ray?
2. Why do Jason's symptoms develop during the season and not during training?
3. Why does the team physician suspect a rotator cuff injury? How do the results of the physical examination support that diagnosis?
4. List the muscles of the rotator cuff.
5. If this injury affected the nerve supply to the muscles of the rotator cuff, which nerve(s) would you suspect, and why?
6. Joyce has broken her femur in a skiing accident and had her leg in a cast for several weeks. When the cast is removed, she is referred to a physical therapist. After having Joyce extend her leg at the knee from a sitting position, the therapist pushes down with her hands on Joyce's shin while asking her to resist the downward force.

Which of the following muscles is being tested?

a. biceps femoris
b. piriformis
c. quadriceps femoris
d. sartorius
e. tensor fasciae latae

7. For some time, Adam has been experiencing fever, lethargy, muscle swelling and tenderness, and a rash on his face, chest, and eyelids. After a number of blood tests and treatments with various anti-inflammatory agents, the symptoms have not been alleviated—in fact, they seem to be worsening. The strength of Adam's upper limb muscles is measured at significantly below normal (3–) on both sides of his body. Based on these signs and symptoms, his physician orders a muscle biopsy. Which muscle disorder does the physician suspect? What questions could the physician ask to aid his diagnosis, and how can the biopsy help identify the disease affecting Adam?
8. A patient has drooping upper eyelids and is unable to raise them. Determine the muscles affected and what nerve(s) may be involved.
9. John, a 25-year-old construction worker, is brought to the hospital emergency room with severe damage to his left forearm. The injury occurred when he caught his arm between a wall and some falling steel beams as part of a building collapsed; it took approximately 2 hours to free him. In the emergency room, his arm appears swollen, and laboratory tests reveal elevated concentrations of serum CK, LDH, and aldolase. As John's condition is monitored, the hospital staff should be alert for

   a. the appearance of dark, reddish-brown urine.
   b. atrophy of the muscles of the left forearm.
   c. contractures in the muscles of the left forearm.
   d. all of the above.
   e. none of the above.

10. Why are muscle pain and stiffness characteristic of trichinosis?

Selected Clinical Terms

compartment syndrome A disorder in which abnormally high pressure in any confined anatomical space impedes blood flow and thus leads to dysfunction of nerves and muscles within the space.

contracture Abnormal shortening of a muscle due to fibrosis, spasm, or other causes, resulting in reduced joint mobility.

crush syndrome A shocklike state occurring when muscles are crushed and the pressure is then relieved; characterized by reduced urine output resulting from damage to the kidney tubules by myoglobin released from the injured muscles.

electromyography The diagnostic recording of electrical activity of the muscles, using either surface electrodes on the skin or needle electrodes inserted into the muscle.

goniometer An instrument for measuring the angle through which an individual can move a joint.

myometer An instrument for measuring the force and extent of muscle contraction.

myopathy Any pathology involving any muscular tissue, but especially referring to the muscular system (skeletal muscle).

myositis Inflammation of a skeletal muscle.

polymyositis Chronic, progressive inflammation of multiple skeletal muscles, characterized by symmetric pain and weakness of the neck, pharynx, and limb muscles, and often accompanied by skin lesions.

rhabdomyolysis An acute, potentially fatal breakdown of skeletal muscle cells characterized by myoglobin in the blood and urine.

rhabdomyoma A benign tumor of either skeletal or cardiac muscle.

rhabdomyosarcoma (RMS) A highly malignant tumor originating in skeletal muscle.

trichinosis Infection with parasitic roundworms of the genus *Trichinella,* typically acquired by eating undercooked pork or other meat; causes myositis when worm larvae invade and encyst in the skeletal muscles.

12 Muscular Tissue

Objectives

In this chapter we will study

- the distinctions between muscular deconditioning and disuse atrophy;
- the neuromuscular disorders botulism, tetanus, and myasthenia gravis;
- the hereditary disease muscular dystrophy;
- the musculoskeletal disorder fibromyalgia; and
- the uses and risks of muscle relaxants.

Diagnosing Functional Disorders of the Muscular System

Chapter 11 described several muscle disorders that are primarily structural. This chapter focuses on the signs, symptoms, and treatment of muscle disorders that are more physiological in nature. It is important to note that the diagnostic techniques described in chapter 11 are equally useful in diagnosing the conditions discussed in this chapter.

Muscle Deconditioning and Disuse Atrophy

Lack of physical activity can cause a muscle to atrophy. **Muscle deconditioning** may become apparent within days to weeks of decreased physical activity. For example, we have all known individuals who were physically fit during their high school days and then became less active over time. The cessation of regular activity causes the muscles to decrease in size, but contrary to what some people say, the muscles themselves do not become "fat." Rather, a continued high level of caloric intake, coupled with decreased activity, leads to increases in fat synthesis and storage by adipose tissue, which shows up most noticeably as subcutaneous fat in various body regions.

Disuse atrophy is a pathological condition in which muscle size is reduced as a result of prolonged inactivity such as bed rest, casting (immobilization of a limb in a cast), or damage to the nerves supplying a muscle (denervation atrophy). Individuals confined to bed have been known to lose up to 3% of their muscle strength per day, and long-term casting can result in the loss of up to 50% of muscle mass.

As a result of disuse, the size of the individual muscle fibers and the oxidative capacity of the mitochondria decline. Muscle fibers can enlarge again if use is restored (for example, when a cast is removed), but regrowth may be compromised if the muscle is not used for more than a year.

Treatments to prevent and/or minimize disuse atrophy focus on moving the immobilized limb, even if in a very limited way. Some of the more common physical therapy treatments are isometric muscle contractions and passive lengthening exercises. In addition, direct electrical stimulation of the immobilized muscles is sometimes used. In this procedure, small surface electrodes are placed on the skin, and a minimal electrical stimulus is applied to the muscle, causing slight contractions that help maintain the muscle.

Neuromuscular Disorders

Some muscular dysfunctions result from disorders of the motor neurons in the nervous system or disorders of the *neuromuscular junction,* the point where a nerve fiber contacts a skeletal muscle fiber. Motor neurons stimulate skeletal muscle fibers to contract by releasing acetylcholine (ACh). Muscle paralysis can result from several types of disorders: inhibition of ACh release (as in botulism); stimulation of excessive ACh release and overstimulation of the muscle (as in tetanus); reduction of the number of ACh receptors on the muscle fibers (as in myasthenia gravis); or blockage of the receptors so that ACh cannot bind to them (the mechanism involved in curare and some other muscle relaxants).

Botulism

Botulism is a form of poisoning caused by the bacterium *Clostridium botulinum,* which is often found in spoiled food but also introduced through skin wounds. *Clostridium* releases the most potent bacterial toxin known—a neurotoxin that prevents motor neurons from releasing ACh. Because these neurons are unable to stimulate the skeletal muscles, the muscles exhibit **flaccid paralysis**—that is, they are limp and unable to contract.

The most common source of botulism in the United States is home-canned food, especially fruit, vegetables, and condiments. The botulism toxin is destroyed by ordinary cooking (boiling for 10 minutes), but the *C. botulinum* spores are highly resistant to heat and require a temperature of 120°C (as in a pressure cooker) to be destroyed. Symptoms of botulism can range from mild digestive upset to severe and fatal reactions. Symptoms typically begin 18 to 36 hours after ingestion; some patients die within 24 hours of exposure. Neuromuscular symptoms typically begin with double vision, difficulty swallowing, and other indications of cranial nerve damage, but then descend to the thoracic muscles where they may lead to paralysis and eventual respiratory arrest.

Patients must be hospitalized and monitored closely. They sometimes require **intubation** (insertion of a nasotracheal tube) or mechanical ventilation. The patient is given antitoxin to neutralize the toxin and antibiotics to eliminate the bacteria. Some patients continue to exhibit recurring symptoms such as weakness and autonomic nervous system dysfunction for as long as a year after exposure.

Tetanus

Tetanus has two meanings in neuromuscular physiology—a state of sustained muscle contraction that is normal and necessary to all muscular action and a pathological muscle paralysis caused by the bacterium *Clostridium tetani. C. tetani* often enters the body through animal bites, punctures, and other unclean skin wounds. It produces a toxin called *tetanospasmin* that causes overstimulation of the muscles. On average, symptoms begin to appear about 7 days after exposure. The muscles exhibit **spastic paralysis,** meaning that they are tense and unable to relax. Often the first effect to be noticed is spasm of the masseter muscles, thus giving tetanus its popular name, "lockjaw." In severe cases, the patient may suffer violent and intensely painful muscle spasms. Since the thoracic muscles must be able to both contract *and* relax in order for a person to breathe, the most serious threat from tetanus is the same as in botulism—respiratory arrest.

C. tetani produces spores that can survive for years in soil and resist disinfectants and boiling. Worldwide, tetanus occurs predominantly in neonates and takes half a million infant lives per year, owing mainly to unsanitary care of the umbilical cord stump. In the United States, however, neonatal tetanus is rare, and the disease has its greatest incidence among elderly people and men whose occupations, such as farming and construction work, put them at risk of soil-contaminated skin wounds. Women are equally susceptible to tetanus, but less often exposed to the risk factors. All people working in the health-care professions have an increased risk of infection, necessitating a conscientious regimen of tetanus vaccination.

A patient with tetanus is kept in a quiet room in intensive care and sheltered from unnecessary stimulation such as noise and light, which can set off violent spasms. Because of the risk of laryngospasm and suffocation, it is important to keep the airway open. An antitoxin is given to bind and neutralize the tetanus toxin, and drugs are used to ease the muscle spasms by various mechanisms. Tetanus can be prevented with a vaccine consisting of an inactivated form of the tetanus toxin; this vaccine stimulates the body to produce antibodies against the toxin. If the toxin is later introduced into the body, these antibodies can incapacitate it before it does any neuromuscular harm. However, periodic booster vaccinations are needed to maintain this immunity.

Myasthenia Gravis

Myasthenia gravis (MG) is an autoimmune disorder of the skeletal muscles. Some further perspectives are added here.

Myasthenia gravis results in a loss of ACh receptors (AChR) from the neuromuscular junction. As a result, nerve fibers become less and less able to excite muscle fibers, leading to muscle weakness and paralysis. Between 70% and 80% of MG patients exhibit pathologies of the thymus, and 15% have thymic tumors. In addition, MG has been linked to other autoimmune disorders such as systemic lupus erythematosus and rheumatoid arthritis (see chapter 10). Of children born to mothers with MG, 10% to 15% exhibit signs of the disease.

MG is divided into three types, depending on the muscles affected. *Generalized myasthenia* involves muscles throughout the body. *Ocular myasthenia* involves the muscles of the eye. *Bulbar myasthenia* involves the muscles innervated by cranial nerves IX, X, XI, and XII. (These nerves are described in chapter 15) Of the three types of MG, bulbar myasthenia progresses most rapidly, and ocular myasthenia occurs most often in males. Generalized myasthenia follows a number of different courses—one characterized by periodic remissions, another that progresses slowly, and a third that progresses rapidly and is called "fulminating."

MG progresses differently in all patients. It typically begins with increased fatigue after exercise and a history of recurring respiratory tract infections. The first muscles that are noticeably affected are those of the head and neck, especially facial and eye muscles. As it progresses to other muscles, the patient may exhibit a drooping or expressionless face, drooling, difficulty swallowing, an altered voice, difficulty holding the head erect, impaired breathing, and coughing. A *myasthenic crisis* occurs when the patient exhibits *quadriparesis* or *quadriplegia*—weakness or paralysis, respectively, of all four limbs. Respiratory paralysis may be not far behind.

In mild cases, spontaneous remission may occur. In others, the disease is characterized by periods of illness followed by intervals of weeks or months that are symptom-free. As MG progresses, the disease periods become longer in duration, and the intervals between them become shorter.

Myasthenia gravis is treated with **cholinesterase inhibitors** and **immunosuppressants.** *Plasmapheresis* is used to remove some of the autoantibodies from the blood. This is a procedure in which blood is removed from the patient and then centrifuged to separate the red blood cells (RBCs) from the plasma. The RBCs are resuspended in physiological saline and returned to the patient's body, while the plasma with the autoantibodies in it is discarded. Because the thymus stimulates the immune system, physicians treating MG may recommend its removal (thymectomy).

Muscular Dystrophy

Muscular dystrophy (MD) is a family of inherited diseases characterized by degeneration of skeletal muscles. Four common forms of MD are Duchenne, limb-girdle, facioscapulohumeral, and myotonic. Table 12.1 shows how these types differ in age at onset, rate of progression, inheritance pattern, and distribution of affected muscles.

The skeletal muscles of affected individuals have a genetic abnormality that is thought to interfere with normal muscle cell metabolism. Affected cells exhibit an increased number of nuclei, which are arranged in chains in the center of the muscle fiber. The muscle fibers undergo necrosis and fragmentation; their myofilaments dissolve. The muscle fibers may be swollen and show irregular striations. Affected muscles exhibit both hypertrophied and atrophied muscle fibers as well as fatty and fibrous infiltration.

MD is diagnosed by a number of methods, including observation of the patient's gait and facial features. Some patients also exhibit tonic muscle spasms. Once a type of muscular dystrophy is suspected, the clinician often uses electromyography (EMG), muscle biopsy, and evaluation of serum enzymes to arrive at a final diagnosis. Analysis of the muscle fibers of patients with Duchenne muscular dystrophy reveals a decreased amount of dystrophin, a protein essential for normal cellular structure and function, and often shows an increased number of abnormal mitochondria. Blood tests show elevated amounts of serum enzymes associated with skeletal muscle.

Each type of muscular dystrophy has a specific treatment regimen. Duchenne muscular dystrophy has no known cure, so the primary goal is to maintain function in unaffected muscle groups and delay the time when the patient must be confined to a wheelchair. Treatment involves range-of-motion exercises, surgical correction of contractures, and special braces. Treatment of facioscapulohumeral and limb-girdle dystrophies includes physical therapy to minimize contractures and orthotic devices for the ankle and foot to help maintain ambulation.

Fibromyalgia

Fibromyalgia (myofascial pain syndrome) is a general name describing a group of musculoskeletal disorders characterized by pain, tenderness, and stiffness in muscles, tendon insertions, and adjacent soft tissues. There is no inflammation. The disease is seen most often in individuals aged 30–50 and in 10 times as many women as men.

Table 12.1 The Four Main Types of Muscular Dystrophy

Type	Age at Onset	Inheritance Pattern	Distribution of Affected Muscles	Rate of Progression
Duchenne	Around 3 years of age	X-linked recessive	Hips, shoulders, quadriceps femoris, gastrocnemius	Rapid
Limb-girdle	Variable	Recessive (not yet established)	Pelvic and shoulder girdles	Variable
Facioscapulohumeral	Between 7 and 20 years of age	Autosomal dominant	Shoulder girdle, neck, face	Moderate
Myotonic	Birth to age 50	Autosomal dominant	Distal extensors, eyelids, face, neck, hands, pharynx	Fast with younger patients; slower with older patients

The exact cause of fibromyalgia is not known, but its etiology appears to be complex. It is often precipitated by another condition, such as chronic fatigue syndrome, Lyme disease, physical or emotional trauma, viral flu-like illness, or HIV infection. Recent studies have reported that altered muscle metabolism (decreased ATP and ADP and higher AMP) and changes in muscle capillary density may be the cellular cause of fibromyalgia. Other studies suggest that the disease is related to functional abnormalities within the central nervous system, since individuals with fibromyalgia may show decreased serotonin and endorphin production and decreased blood flow in the thalamus, one of the brain regions involved in pain perception.

Fibromyalgia is diagnosed when a patient complains of chronic diffuse pain with a gradual onset that is aggravated by straining and overuse of muscles. The pain is localized to one of nine pairs of tender points: the occiput, trapezius, supraspinatus, gluteal region, greater trochanter, knee, lateral epicondyle, second rib, and lower cervical spine. The pain is often accompanied by a number of other symptoms, such as fatigue, irritable bowel, insomnia, anxiety, headache, and cold sensitivity. In making a diagnosis, it is important to exclude other diseases, such as arthritis, polymyositis, polymyalgia, and connective tissue disorders. Fibromyalgia patients show normal results in EMG, muscle biopsy, and blood chemistry testing.

Once diagnosed, fibromyalgia is treated with a combination of techniques that includes stretching exercises, therapeutic massage, local applications of heat, and aspirin or other nonsteroidal anti-inflammatory drugs (NSAIDS). However, it is important to note that a single treatment regimen able to relieve symptoms in all patients has not been established. Milder cases of fibromyalgia have been observed to remit spontaneously if stress is decreased, but they can also become chronic and recur at more frequent intervals if stress increases or remains unchanged. Education and reassurance are believed to be extremely important in treating fibromyalgia.

The Clinical Use of Muscle Relaxants

Muscle relaxants are drugs used to inhibit skeletal muscle contraction. They are employed for such purposes as treating spastic muscle contraction and relaxing the abdominal muscles for surgery. They allow a lighter level of general anesthesia, since muscle relaxation is no longer dependent on the anesthetic. Lighter anesthesia reduces the risk of cardiovascular and respiratory depression.

Curare is a naturally occurring muscle relaxant derived from vines of the genus *Strychnos* and other tropical plants. It was first used clinically in 1932 to treat the muscle spasms of tetanus and other neuromuscular disorders. It was first used in surgery in 1942. Curare acts by blocking ACh receptors, but most muscle relaxants used now act at the level of the central nervous system—for example, by enhancing the activity of neurotransmitters that inhibit motor neurons.

Muscle relaxants must be used with great care, because overdoses can cause side effects ranging from unpleasant (vomiting, double vision, and hallucinations) to life-threatening (cardiac arrhythmia, seizures, and hypotension). They are usually administered by anesthesiologists or others specially trained in their use, and only in clinical settings where cardiovascular and respiratory resuscitation equipment is close at hand if needed.

Case Study 12 The Professor with Double Vision

Max, a 55-year-old college professor, visits his physician for a routine physical. Like most people his age, he has had his share of colds, sprained ankles, scrapes, cuts, and bruises, but no previous serious injury or illness. Max has always been physically active and prides himself on being in good shape for his age. Lately, however, he becomes fatigued more easily than before. He also reports some difficulty chewing and swallowing, and says, “Sometimes I start seeing double when I’m watching the late news on TV.”

While taking Max’s history, the physician notes that he speaks in a somewhat slurred, high-pitched monotone. The doctor asks Max to blink his eyes five times at 2-second intervals. He notices that Max takes longer to reopen his eyes each time, and after the third blink, he keeps his eyes closed for about 15 seconds, almost as if he has fallen asleep. From Max’s medical record, the doctor notes that Max’s mother suffered from rheumatoid arthritis and his father had a thyroid disorder. Results of Max’s physical examination are shown here.

Mental status: Alert

Weight: 172 lb; has lost 10 lb since routine physical 12 months ago

Oral temperature: 98.6°F (37.0°C)

Heart rate: 68 beats/min

Respiratory rate: 11 breaths/min

Blood pressure: 140/85 mmHg

Sensory function: Cannot sustain straight gaze; left eye wanders after 10–15 seconds, and patient reports seeing double.

Movement and reflexes: Voluntary movement somewhat sluggish; some weakness noted in arms; patellar tendon reflex normal.

A blood test reveals the presence of serum antibodies against the ACh receptor, leading to a diagnosis of early myasthenia gravis. The physician then refers Max for a clinical test with a drug called *edrophonium.* Edrophonium is a cholinesterase inhibitor. When given intravenously, it produces a brief (< 5 min) improvement in muscular function. It must be given in a clinical setting under careful observation, because in conditions other than MG, edrophonium can cause a dangerous depression in cardiopulmonary function.

The physician tells Max that there is no cure for MG, but that the symptoms are treatable. He says MG is a difficult disease to manage, so he refers Max to a specialist who prescribes *neostigmine.* This is an oral cholinesterase inhibitor that prolongs the action of ACh at the neuromuscular junction by slowing its breakdown by AChE. In order to reduce the autoimmune attack on Max’s neuromuscular junctions, the specialist recommends three measures: (1) *thymectomy;* (2) *plasmapheresis;* and (3) *prednisone,* an immunosuppressive corticosteroid.

The specialist tells Max that he may experience periods of remission, but his symptoms will increase in severity over time. Possible complications include extreme muscle weakness and respiratory difficulties, most likely accompanied by respiratory infections.

Based on this case study and other information in this chapter, answer the following questions.

1. Why do you think Max’s voice is altered by his condition?

2. Many myasthenia gravis patients have difficulty holding their heads erect. Explain.

3. Why do you think Max takes longer and longer to reopen his eyes each time he blinks? In a related vein, why do you think he notices double vision especially when watching the late-night news, rather than at other times of day?

4. Max's doctor remembers a previous myasthenia gravis patient who said she often woke up in the middle of the night choking on her saliva. Explain why that would occur.

5. A thymectomy will not be done, of course, without Max's informed consent. The specialist advises Max that, although the benefits of thymectomy outweigh the risks in cases like his, it does carry a risk of increased susceptibility to infectious diseases such as pneumonia. Explain why this would be true.

6. Curare has been known for centuries to the natives of South America, who extract it from a vine and apply it to the tips of the blowgun darts they use for shooting down monkeys and parrots from the high jungle canopy. Explain why curare would be useful for this purpose.

7. Susan, a 16-year-old high school distance runner, is training with her team when she steps in a hole in the ground and feels a "pop" in her left foot. She tries to continue running, but it hurts too much. Suspecting that she has sprained her ankle, her coach refers her to the team doctor, who orders an X ray. To her surprise, Susan learns that she has fractured the third metatarsal bone and will have to wear a cast on her foot for 6 weeks. When the cast is removed, Susan notices that her left calf is markedly smaller than her right. Why did this occur, and how could the size difference have been minimized?

8. Jason, a 3-year-old, explores under the kitchen sink at his grandparents' house and ingests some ant poison. The warning label on the package says that the active ingredient in the poison is a cholinesterase inhibitor. At the hospital, Jason's treatment is similar in some ways to the way a tetanus patient would be treated. Predict how the two treatments would be similar and explain why.

9. Carla, a 60-year-old public health nurse, is herself under treatment for myasthenia gravis. Her profession calls for periodic tetanus boosters. Explain why the effectiveness of these booster shots may be compromised by her treatment regimen for MG.

10. Rhonda, who has experienced fatigue and muscle pain for 6 years, has finally been diagnosed with fibromyalgia. She complains to her friend about how long it took to diagnose her condition and accuses the doctors of not knowing what they're doing. If you were to take the doctors' side, what would you give as some reasons that fibromyalgia is so hard to diagnose?

Selected Clinical Terms

cholinesterase inhibitor A chemical that inhibits the action of acetylcholinesterase, thus slowing the rate of acetylcholine (ACh) breakdown and prolonging or intensifying the action of ACh at a synapse.

disuse atrophy A loss of muscular mass and strength as a result of the immobilization of a body region or damage to the nerve that innervates a muscle.

flaccid paralysis A state in which a muscle is relaxed (flaccid) and unable to contract.

immunosuppressant A drug that inhibits the immune system; used for such purposes as treating autoimmune diseases or preventing the immune rejection of a transplanted organ

intubation Insertion of a tube in order to maintain an open (patent) passage such as the airway.

muscle deconditioning A loss of muscular mass and strength as a result of reduced exercise.

spastic paralysis A state in which a muscle is contracted and unable to relax.

13 Nervous Tissue

Objectives

In this chapter we will study

- methods for diagnosing nervous system disorders;
- two demyelinating diseases—Guillain-Barré syndrome and multiple sclerosis;
- the characteristics and treatment of brain tumors, especially gliomas; and
- a brief overview of stroke, or cerebrovascular accident.

Diagnosing Nervous System Disorders

The nervous system facilitates communication among the trillions of body cells through rapid electrical conduction of signals. When this system is disrupted, many other body systems are affected as well. The diagnosis of neurological disorders is especially challenging, not only because the nervous system itself is so complex, but also because the signs and symptoms of neurological disorders can be difficult to distinguish from those of other organ systems.

A neurological assessment is normally incorporated into a standard physical examination. Functions that the examining physician routinely tests include the following:

- **special senses**—vision, hearing, smell, taste, and balance;
- **general senses**—touch, temperature sensation, vibration, and pain;
- **motor function**—eye movements, tongue movements, voice, swallowing, facial expressions, muscle strength, symmetry of musculature, signs of muscular atrophy, body position, gait, and coordination;
- **deep tendon reflexes**—muscular responses elicited by striking a tendon with a reflex hammer in such sites as the knee, elbow, wrist, ankle, and heel, or by using the handle of the hammer to scratch the skin in such sites as the abdomen and sole; and
- **level of consciousness**—states of responsiveness, or lack thereof, such as alertness, lethargy, or stupor. There are specific definitions and tests for these and other states.

When nervous system disorders are suspected, a more complete neurological examination is conducted. As with any other disease, the patient history provides the first clues. For example, the patient's social and travel history may indicate possible exposure to toxins or infectious agents that could affect the nervous system, while the family history may suggest inherited neurological disorders. Some of the most common neurological symptoms reported by patients are dizziness, insomnia, fatigue, back pain, headache, muscle weakness, and **paresthesia**—abnormal sensations such as prickling, burning, or tingling. Once a neurological disorder is suspected, the clinician must determine the site of the abnormality, such as the brain, the spinal cord, a nerve, or a muscle.

A complete neurological examination evaluates mental status, cranial nerve function, sensory function, reflex function, autonomic function, and the cerebral vasculature. In addition to the basic imaging techniques described in chapter 2, some specialized imaging methods for the nervous system include:

- **cerebral angiography,** in which the cerebral blood vessels are examined by injecting an opaque dye into the circulatory system and taking an X ray of the head;
- **myelography,** in which dye is injected into the nervous system itself to render the spinal cord and nerve roots visible on X rays; and
- **echoencephalography,** ultrasound examination of the nervous system.

In addition to imaging techniques, brain function can be measured using either an *electroencephalogram,* which is a recording of the spontaneous electrical potentials of the brain, or *evoked potentials*, which are electrical responses of the brain to specific external stimuli. In a *lumbar puncture (spinal tap),* a sample of cerebrospinal fluid (CSF) is aspirated and examined for color, composition, and the presence of blood cells or microorganisms. The normal appearance and constituents of the CSF can be found in the Appendix of Normal Values at the end of this manual.

Demyelinating Diseases

The myelin sheath found around many nerve fibers is important to their conduction speed and the precision of conduction pathways. Disorders of the sheath can thus disrupt nerve conduction and cause a variety of sensory, motor, and cognitive dysfunctions. Two examples of this are Guillain-Barré syndrome and multiple sclerosis.

Guillain-Barre Syndrome

The most common demyelinating disorder is Guillain-Barré (gee-YAN buh-RAY) syndrome (GBS), also known as *acute idiopathic polyneuropathy* and by several other names. GBS is a pathology of the peripheral nerves that affects people of both sexes and all ages. It typically follows recovery from a viral infection, but in rare cases it may be triggered by lymphoma, surgery, or a vaccination against influenza or some other viral disease. GBS is thought to be an autoimmune disease in which autoantibodies attack the peripheral nerves, the myelin degenerates, and the nerve fibers themselves sometimes follow. When it affects motor nerves, GBS results in muscle weakness and atrophy, although motor function can recover if the cell bodies of the neurons survive and the nerves regenerate.

GBS is characterized by acute-onset muscle weakness that typically progresses from the lower to the upper limbs. The first sign is a relatively symmetrical weakness in both legs, often accompanied by paresthesia. Muscle tone declines, and deep tendon reflexes are lost. As GBS progresses, the upper limbs are affected, and then the muscles of the face and neck. GBS also affects the autonomic nervous system, which controls much of our visceral function. This results in an elevated heart rate, unstable blood pressure, and shortness of breath. In severe cases, death sometimes results from respiratory paralysis and autonomic dysfunctions.

GBS is diagnosed from the patient history, examination of the CSF, electromyography, and nerve conduction tests. The CSF of a GBS patient contains abnormally high protein concentrations (500 mg/dL), but no abnormal cells. Nerve conduction speed is decreased.

The treatment of GBS depends upon the severity of the condition. Less severe cases can be treated with plasmapheresis to remove the autoantibodies from the blood (see chapter 12 of this manual). This shortens the course of the disease and reduces the risk of permanent paralysis or death. Heat is used to relieve the pain of GBS and thus make it possible for the patient to undergo other physical therapy. A physical therapist initiates passive range-of-motion exercises early in the treatment and encourages active exercise as the patient regains mobility. In severe cases, the patient is usually hospitalized to monitor vital functions and provide respiratory support if needed. These patients are susceptible to pressure ulcers and fluid imbalances. Caregivers must passively exercise the patient's joints to prevent *ankylosis* (see chapter 10 of this manual). About 30% of adults who recover from GBS continue to show some muscular weakness even 3 years later. The residual effects of GBS may require occupational or physical therapy or orthotic appliances to provide the patient with some degree of independence.

Multiple Sclerosis

Multiple sclerosis (MS) usually strikes between the ages of 20 and 40 years and affects women twice as often as men. Like GBS, it apparently results from an autoimmune attack, often triggered by a viral infection. MS, however, strikes the oligodendrocytes—myelin-producing cells of the CNS—and does not affect the nerves of the PNS. The white matter of the CNS exhibits widespread plaques of scar tissue, often surrounded by regions of edema. As demyelination and scarring spread, nerve conduction becomes progressively impaired, and the signs and symptoms worsen.

The exact cause of MS is unknown, but it involves both hereditary and environmental risk factors. It is about nine times more likely to occur in two monozygotic twins (genetically identical twins who develop from the same fertilized egg) as in two dizygotic twins (nonidentical twins who develop from separate fertilized eggs). Whites are more susceptible to MS than other groups of people, and about 15% of people with MS have relatives with the same disease. MS is five times more common in people who live their first 15 years in temperate climates than in people of the tropics; moving after the age of 15 does not affect the risk level. Thus, it appears that MS results from a combination of heredity and an environmental factor, possibly a virus, encountered in childhood. Conditions such as trauma, infection, and fatigue often seem to set off the initial attack of MS in people who have these risk factors. The fatigue following pregnancy is sometimes linked to the onset of MS.

There are multiple types of MS: a *mixed type* characterized by rapid onset of visual defects; a *spinal type* characterized by weakness or numbness in the limbs and loss of bladder control; and a *cerebellar type* that involves a loss of coordination and a spastic gait. About half of MS patients show a mixture of these types after several years. Some cases of MS exhibit alternating **remission** (disappearance of signs) and **relapses** (reappearance of signs). Victims may exhibit several **paroxysms** (attacks) each day—episodes of paresthesia, **ataxia** (lack of coordinated movement), and **dysarthria** (speech disturbances) lasting from a few seconds to minutes.

Diagnosis of MS is based on the patient history and physical examination, with those findings confirmed by other tests. CSF immunoglobulin concentrations are found to be elevated in approximately 67% of MS patients. Imaging techniques such as MRI and CT scans are used to identify the location and severity of the plaques. MRI is the most sensitive diagnostic technique, often revealing plaques overlooked by CT. If these methods fail to indicate exact locations, evoked potential studies can be used to confirm the diagnosis.

Over the course of the disease, the patient experiences symptoms that require both supportive and rehabilitative therapies. These symptoms include weakness, pain, spasticity, depression, heat intolerance, tremor and ataxia, and bladder, bowel, and sexual dysfunction. Muscle training provides both physical and psychological benefits. The patient should be encouraged to maintain as normal and active a lifestyle as possible, but overwork and fatigue must be avoided. As the patient gets weaker, he or she becomes more bedridden, but this result can be delayed by physical therapy and prompt treatment of infections and urinary difficulties. In bedridden patients, preventing pressure ulcers and infections of the lungs and urinary tract is of utmost concern.

At present, there is no cure for MS. The average duration of the disease is about 30 years; many patients live a normal life span, while some die within a year of onset. Short-term corticosteroid therapy sometimes shortens acute attacks of MS and may reduce the risk of long-term neurological deficits. Subcutaneous injections of interferon reduce the frequency of relapses for some patients.

Gliomas

Brain tumors are classified as **primary tumors** if they arise from brain tissue itself and **secondary (metastatic) tumors** if they have invaded the brain by metastasis from a tumor elsewhere in the body (for example, metastatic lung or colon cancer that has invaded the brain). Primary brain tumors are further classified as *extracerebral* or *intracerebral.* **Extracerebral tumors** arise from structures outside the brain, such as the pituitary or pineal gland, cranial nerves, or meninges (membranes around the brain); **intracerebral tumors** arise from the brain tissue itself. Primary intracerebral tumors are also called **gliomas** because they usually arise from glial cells.

Some gliomas are encapsulated and noninvasive—that is, they are confined by a fibrous capsule and do not spread easily to adjacent neural tissue. They can nevertheless be life-threatening because they compress and displace other CNS tissue and cerebral blood vessels. Other gliomas are nonencapsulated and invasive; these tumors spread into adjacent tissue and destroy it.

Gliomas can arise from almost any type of CNS glial cell—for example, *astrocytomas* develop from astrocytes, *oligodendrocytomas* from oligodendrocytes, and *ependymomas* from ependymal cells. The signs and symptoms of a glioma depend on its location and type. Headaches may be an early symptom, sometimes accompanied by irritability and personality changes. As the tumors multiply and grow, they create increased intracranial pressure, often producing such signs as vomiting, seizures, **vertigo,** paresthesia, and motor and sensory deficits.

Treatment of a glioma depends on the type, location, and severity of the tumor. If at all possible, the tumor is surgically removed. Surgical excision is often coupled with either chemotherapy, radiation therapy, or both. If surgical removal is not feasible due to the tumor's location, chemotherapy and radiation therapy are usually used. However, chemotherapy for brain tumors is often hindered by the *blood-brain barrier,* a system that makes it difficult to get drugs into the CNS tissue.

Stroke

More than 50% of hospital admissions for neurological problems are due to **stroke,** or **cerebrovascular accident**—a sudden loss of blood supply to the brain. Strokes occur either because a cerebral artery is obstructed (often by a fatty atherosclerotic deposit, a blood clot, or both) or because a weakened cerebral artery has hemorrhaged. These pathologies are usually secondary to hypertension (elevated blood pressure), atherosclerosis, or a combination of the two pathologies. Death (infarction) of brain tissue may or may not occur. The neurological effects of a stroke vary widely, depending on what region and how much brain tissue is damaged; blindness, speech defects, and paralysis are a few common consequences. Stroke is further discussed with other vascular diseases in chapter 20 of this manual.

Case Study 13 The Drummer with Tingling Fingers

Aaron, a 26-year-old musician, visits his physician complaining of tingling in the fingers of his right hand. The feeling is present when he plays his drums as well as at other times of the day and night. Sometimes the tingling is so bad that he has difficulty feeling anything with his right hand and ends up dropping things. He has also noted that his right hand and arm get tired more easily than his left hand. In addition, he has had problems seeing correctly for the past 3 weeks; even during the day or in bright rooms, his overall vision is "darker" than normal. At times he feels like something is crawling over the right side of his face. Finally, Aaron mentions that during the time he has been most worried about these symptoms, his legs have felt weak and he has been tripping over things.

Examination reveals weakness of the rectus muscles of Aaron's right eye and mild weakness of his right facial muscles. Other muscles are of normal strength. Aaron exhibits normal reflexes, but his right-side reflexes are somewhat greater than those on his left side. The physician suggests that Aaron get more rest and have his eyes checked because he may need glasses. The physician also tells him to return if his condition does not improve.

Three months later, Aaron comes back. In addition to his previous symptoms, he has developed difficulty walking and speaking. Although he frequently feels the need to urinate, he is unable to fully empty his bladder. On this visit, the physical examination shows disturbances in Aaron's gait—he has become ataxic, and his stance is wider than normal. His superficial reflexes are diminished, and his deep tendon reflexes are exaggerated. Based on these signs, the physician orders MRI scans and a spinal tap. The MRI results show areas of demyelination and plaques in the white matter of the brain. When the CSF is analyzed, elevated concentrations of leukocytes, protein, and antibodies are found, and myelin basic protein is present. These results lead to a diagnosis of multiple sclerosis.

Based on this case study and other information in this chapter, answer the following questions.

1. Why is multiple sclerosis not diagnosed initially?
2. How do Aaron's physical signs and symptoms support the diagnosis of multiple sclerosis? How could you rule out Guillain-Barré syndrome?
3. What treatments would you expect the physician to prescribe?
4. If you were Aaron's physical therapist and he asked your opinion of his prognosis, what would you tell him?
5. Why do plaques appear in the CNS of a patient with multiple sclerosis?
6. Suppose someone argued that the reason twins often share multiple sclerosis is not because it's hereditary, but because of some abnormality in the mother's pregnancy such as maternal exposure to a toxin or virus. What argument could you give against this hypothesis?
7. Autoantibodies are present in both Guillain-Barré syndrome and multiple sclerosis. Plasmapheresis is helpful in treating the former but not the latter. Explain this difference.

8. Are all gliomas malignant? Explain your answer.
9. Why are seizures a characteristic sign of gliomas?
10. Why are glial cells more likely than neurons to produce primary intracerebral tumors?

Selected Clinical Terms

ataxia An inability to coordinate voluntary muscular activity of the limbs, trunk, or head; often suggests lesions of the cerebellum or spinal cord.

dysarthria A disturbance of speech, often resulting from brain lesions or from paralysis or spasticity of the muscles of speech.

paresthesia Abnormal sensations such as tingling, burning, or prickling, often in the absence of external stimulation.

paroxysm Any sudden onset or "attack" of a disease or intensification of symptoms; a sharp spasm or seizure.

primary tumor A tumor that originates from cells in the organ where it is found.

relapse A recurrence of the symptoms of a disease following a period of remission or improvement in the person's condition.

secondary tumor A tumor that did not originate in the organ where it is found, but arrived there by metastasis from a primary tumor elsewhere; also called a metastatic tumor.

vertigo A sensation of spinning or whirling, or an illusion of objects moving around the person, often indicating disorders of the inner ear or brain; often used less precisely to denote a sensation of dizziness or lightheadedness.

14 The Central Nervous System

Objectives

In this chapter we will study

- two spinal cord diseases—poliomyelitis and amyotrophic lateral sclerosis;
- two inflammatory brain diseases—meningitis and encephalitis;
- two types of traumatic brain injury—concussion and contusion; and
- the pathogenesis and types of epilepsy.

Poliomyelitis

Poliomyelitis, also known as *polio* or *infantile paralysis,* is an inflammation of the gray matter of the spinal cord that sometimes leaves a person with permanent paralysis. It primarily strikes children. Polio was prevalent in the United States in the 1950s (peaking at 21,000 cases in 1952), but is now nearly nonexistent because of vaccination. The 5 or 10 cases that occur per year are side effects of the vaccine.

Polio is caused by a relatively small virus that usually enters the body orally and multiplies in the tissues of the mouth and intestines. The virus next invades the blood, replicates again, and then enters the CNS. Even though the virus resides in numerous tissues at various times, it affects only motor neurons. Neuronal damage results in inflammation of the CNS and eventual destruction of neurons by phagocytic cells. The severity and site of the resulting paralysis are determined by the specific motor neurons destroyed.

Most cases of poliovirus infection are asymptomatic. In severe cases, however, the victim exhibits fever and severe back and neck pain followed by progressive muscle weakness and ultimately paralysis, especially in the lower limbs. *Bulbar paralysis,* caused by lesions of the brainstem, affects the oral and pharyngeal muscles and thus causes difficulties with speech, swallowing, and aspiration of secretions.

Polio is incurable, and the treatment is palliative. Analgesics and antipyretics are used to relieve the pain and fever, respectively. Muscle spasm and pain are treated with hot packs. Fewer than 25% of victims suffer permanent paralysis, and more than half recover completely. If paralysis occurs, however, physical therapy is employed to restore as much mobility and independence as possible. Respiratory therapy and artificial ventilation are sometimes required for cases that involve the neurons of respiration.

Amyotrophic Lateral Sclerosis

Amyotrophic lateral sclerosis (ALS), also known as *Lou Gehrig Disease,* is a degenerative disease of the motor neurons. The signs of ALS usually begin to appear after the age of 30; the median age at onset is 55. ALS affects about one in 100,000 persons. Prior to menopause, women are affected at only two-thirds of the male rate, but after menopause, the incidence of ALS in women is about equal to that in men. Some people with ALS (about 2.5%) exhibit a defect in the gene for superoxide dismutase (SOD), an enzyme that neutralizes the superoxide free radical. This suggests that at least some forms of ALS result from free-radical damage to neurons. Other studies have implicated abnormal glutamic acid metabolism and hydrogen peroxide production as causes.

ALS is characterized by the noninflammatory degeneration of upper and lower motor neurons. As motor nerve fibers degenerate, glial cells proliferate and form scars in the corticospinal tract of the spinal cord. Adjacent neurons attempt to compensate for the loss by sprouting new processes and reinnervating the muscles "abandoned" by the dying neurons. Thus, these compensating neurons take on control of larger and larger motor units. As they do so, their control of muscle movement becomes less and less precise.

The characteristic effects of ALS include:

1. **paresis** (muscle weakness) beginning in a single muscle group;
2. asymmetric involvement of muscles on the right and left sides of the body;

3. gradual progression to all muscles of the body except for the muscles of eye movement;
4. both flaccid and spastic paralysis, often appearing within a single muscle group; and
5. an absence of autonomic, sensory, or mental effects; these functions remain normal.

ALS is diagnosed mainly from the patient history and physical examination. Electromyography and muscle biopsy may confirm lower motor neuron degeneration and denervation, but are not necessary to the diagnosis. ALS is incurable. Treatment is aimed at maintaining the patient's ability to communicate, managing the dehydration and weight loss that result from difficulty chewing and swallowing, and managing the respiratory difficulty *(dyspnea)* that results from weakness of the diaphragm and intercostal muscles. The average life expectancy after the onset of symptoms is 2 to 3 years, but some patients survive for 15 years or longer.

Meningitis

Meningitis is inflammation of the meninges of the brain, spinal cord, or both, resulting from a toxin or infection. Meningitis is classified according to its etiology—for example, bacterial meningitis or viral meningitis. It can also be classified as acute, subacute, or chronic, depending on the duration of the inflammation. In general, acute meningitis has a relatively short duration (less than 2 weeks), subacute meningitis lasts 2 to 4 weeks, and chronic meningitis lasts for more than a month.

Bacterial meningitis has the highest incidence among infants, especially in the first month of life. It affects about 20 in 100,000 full-term infants and 10 times as many infants who are underweight at birth. Infantile meningitis is usually due to *Hemophilus influenzae* or *Escherichia coli.* Childhood and adult meningitis is usually caused by *Neisseria meningitidis* (meningococcus) and *Streptococcus pneumoniae* (pneumococcus). The second-highest incidence is among students living in college dormitories, where the incidence is about 4.6 per 100,000. About 100 to 125 college students in the United States contract bacterial meningitis annually, with 5 to 15 deaths and 5 to 15 more who suffer deafness, brain damage, or loss of limbs as a result of infection. The CDC urges that college students, especially dormitory residents, be vaccinated and that they report to a doctor immediately if they experience a combined headache, fever, and stiff neck.

Meningococcus and pneumococcus are common inhabitants of the pharynx. Upper respiratory infections can enable these bacteria to travel in the blood and invade the CNS. CNS infections involve the pia mater, arachnoid, subarachnoid space, ventricles, and CSF. The bacteria or their toxins cause inflammation in the meninges and ventricles and make the meningeal blood vessels leaky. Blood cells (especially neutrophils) then get into the subarachnoid space and produce an exudate that thickens the CSF and damages the cranial nerves. The meninges become edematous, intracranial pressure rises, the blood vessels and choroid plexuses become inflamed, and astrocytes and microglia proliferate in the nervous tissue.

The signs and symptoms of acute bacterial meningitis include a severe, throbbing headache; fever; neck stiffness and flexion of the neck toward the chest; taut flexion of the hips and knees; and projectile vomiting. The state of consciousness declines from irritability and confusion to reduced responsiveness and drowsiness, followed by delirium and coma. The victim can die from massive brain *infarction* or from systemic problems such as septic shock. Adults can become critically ill within 24 hours of the first symptoms, and infants even more quickly. Acute bacterial meningitis can be fatal in a matter of hours, so in cases where the disease is suspected, the patient is often treated with antibiotics even before the test results have been received.

Bacterial meningitis is diagnosed primarily by lumbar puncture (spinal tap) and examination and culture of the CSF. The skin, head, and ears are also closely inspected to determine if there has been a puncture wound, insect bite, or some other source of entry for a bacterium. Once the diagnosis is confirmed, the patient is given antibiotics specific to the bacterial type, and supportive therapy is continued for the fever and for the dehydration and electrolyte imbalances resulting from vomiting.

Viral meningitis is usually confined to the meninges. It is caused by a wide variety of viruses, including but not limited to arboviruses, poliovirus, and Coxsackievirus. The arboviruses are most commonly transmitted by mosquitoes. Viral meningitis is most often acute, with varying signs and symptoms. Some patients may be asymptomatic; others exhibit fever and malaise with no CNS effects even in the presence of abnormal CSF, or show the severe symptoms of bacterial meningitis.

Diagnosis of viral meningitis is based on the characteristics of the CSF, including normal glucose concentrations and the absence of bacteria. Diagnosis is most often accomplished by comparing the serum antibodies from infected patients with existing serum samples from previous patients. Treatment is aimed at supportive therapy and prevention of viral replication.

Fungal meningitis is a less common and more chronic disease than bacterial or viral meningitis. It is often seen in individuals with impaired immune function. The most common causative fungi are *Cryptococcus, Coccidioides, Mucor, Candida, Actinomyces, Histoplasma,* and *Aspergillus.* The incidence of fungal meningitis has increased with the use of immunosuppressive therapies and the incidence of AIDS. The signs and symptoms are similar to those of bacterial and viral meningitis, but they develop much more slowly. Diagnosis is again through lumbar puncture and evaluation of CSF. In this case, the CSF exhibits decreased glucose and elevated protein concentrations, and the causative agent can be cultured. After the organisms have been identified, the appropriate drug is chosen to treat the infection. Once again, supportive therapy is provided to alleviate fever, dehydration, and electrolyte imbalance.

Encephalitis

Encephalitis is a **febrile** inflammation of the brain—that is, accompanied by fever. Meningitis does not always lead to encephalitis, but when encephalitis exists, it always coexists with meningitis; the combination of disorders is called *meningoencephalitis*. Encephalitis is usually caused by mosquito-borne arboviruses and the herpes simplex virus, but it can also occur as a complication of other viral infections such as polio, rabies, and mumps; typhus (a bacterial disease); or parasitic diseases such as malaria and some worm infections. Some cases are mild and result in full recovery, whereas others involve extensive edema, hemorrhaging of the brain, and necrosis of the nervous tissue. These conditions can lead to delirium, confusion, seizures, unconsciousness, and death. Encephalitis is diagnosed from the patient history, physical examination, CSF examination, WBC count, and MRI and CT scans. The disease is difficult to treat, although some progress has been made with antiviral drugs such as acyclovir.

Traumatic Brain Injury

Traumatic brain injury (TBI), or **head trauma,** occurs with greatest frequency among infants, young school-age children, people between 15 and-30 years of age, and the elderly. It causes more death and disability than any other neurological disorder among people under 50; it is three times as common among men as among women; and it is a leading cause of death among boys and men under the age of 35. Some common causes of TBI include motor vehicle accidents, sports injuries, falls, and assault.

Traumatic brain damage is classified as *open (penetrating* or *missile) trauma* if the dura mater is breached by a penetrating wound and brain tissue is exposed, or *closed (blunt* or *nonmissile) trauma* if the brain is not exposed. Open trauma, usually caused by bullets or sharp projectiles such as knives, can cause cerebrospinal fluid to leak through the nose or ear and can introduce infectious organisms to the CNS. Closed trauma, however, is more common. It frequently results when the head strikes a hard surface or is hit by a rapidly moving object. Damage to the nervous tissue, meninges, and blood vessels often occurs when the brain strikes the inside of the cranium or when the base of the brain shears across the bones of the cranial floor. For example, when the head strikes the dashboard in an automobile accident, the brain strikes the front of the cranium and then rebounds off the rear, causing injury in two places: *coup* injury to the frontal lobes and *contrecoup* (con-treh-COO) injury to the occipital lobes.

Injury to the CNS can be primary, secondary, or tertiary. *Primary injury* is trauma at the site of impact. *Secondary injury* affects a wider zone of brain tissue as a result of cerebral edema, hemorrhage, ischemia, infection, or increased intracranial pressure. *Tertiary injury* results from altered neural regulation of other body systems. It can involve **apnea** (temporary cessation of breathing), **hypotension** (low blood pressure), and alterations in cardiac function.

The signs and symptoms of TBI depend on the regions of the CNS affected. Diagnosis is based on the patient history, physical examination, and often medical imaging. In addition, various clinical tests can help to either confirm TBI or exclude other disorders. For example, if intracerebral bleeding is suspected, a lumbar puncture may be performed to determine whether blood cells are present in the CSF. Signs that indicate damage to the brainstem include coma, irregular breathing, unresponsive pupils, loss of visual reflexes, loss of balance, and muscle flaccidity. Severe head trauma has a 50% mortality rate, and treatment reduces this rate only slightly.

Although a wide variety of TBIs exist, the following discussion focuses on concussion and contusion.

Concussion

Even though head injuries are a leading cause of death, 75% to 90% of them are not severe. **Concussion,** the most common brain injury, is caused by blunt trauma and rarely involves serious neurological impairment. Concussions are graded from I to IV based on their signs. Grades I to III involve increasing degrees of confusion and memory loss (especially the memory of events that occurred just before the injury), but no loss of consciousness. Grade IV *(classic cerebral concussion)* disrupts communication between the cerebral cortex and the reticular formation of the brainstem, and thus results in loss of consciousness lasting for as long as 6 hours. More severe trauma causes **diffuse axonal injury (DAI),** in which nerve fibers are damaged and permanent neurological deficits may occur.

Contusion

A brain **contusion** is a more severe injury that breaks blood vessels and causes hemorrhaging. This usually occurs on the surface of the brain, but it can also involve the accumulation of a massive **hematoma** (blood clot) that kills nervous tissue by compression. A blood clot between the cranial bone and dura mater is called an *epidural hematoma,* and a clot deep to the dura mater is a *subdural hematoma.* The latter is more common. Hematomas can be found by CT and MRI scanning. If one exists, it is important to evacuate it before it causes death of brain tissue.

Contusions typically accompany severe surface injuries and skull fractures, and frequently produce **hemiplegia** (paralysis of one side of the body) and other signs of focal cortical dysfunction. Muscle rigidity, respiratory irregularity, unreactive pupils, and coma often occur in severe cases. Epidural and subdural hematomas often produce delayed neurological effects as nervous tissue is gradually destroyed by the increased intracranial pressure. The signs and symptoms of hematoma may not appear until weeks after the trauma. They can begin with symptoms as mild as headache and drowsiness, but progress to **hemiparesis** (unilateral muscle weakness), spasticity, and finally, deepening coma and death.

Epilepsy

Epilepsy is a recurrent disorder of cerebral function in which excessive discharge of cerebral neurons causes sudden, brief attacks of muscle spasticity, sensory hallucinations, altered consciousness, or inappropriate behavior. Epilepsy is classified as *symptomatic* if it can be linked to an identifiable cause and *idiopathic* if it cannot. About 75% of cases among young adults are idiopathic. Symptomatic epilepsy can be a result of developmental (prenatal) abnormalities, birth trauma, hereditary predisposition, brain infections, tumors, vascular diseases, alcoholism, drug abuse, or traumatic brain injury.

Many cases of epilepsy are misclassified as idiopathic until, after death, an autopsy reveals a cause, such as a microscopic scar produced by birth trauma. Seizures that begin before the age of 2 years are usually due to birth trauma, developmental defects, or inherited metabolic diseases. Seizures that begin after age 25 are usually due to traumatic brain injury, tumors, or other brain diseases. However, seizures can begin at any age as a result of localized brain lesions. About 10% of severe closed TBIs and 40% of open TBIs are followed by **posttraumatic epilepsy,** with seizures often beginning several years after the trauma.

There are many kinds of epileptic seizures. Two of the best known are *absence attacks (petit mal seizures)* and *tonic-clonic (grand mal) seizures.* An **absence attack** is a brief (10–30 sec) loss of consciousness, with fluttering of the eyes or muscles and sometimes a loss of muscle tone. The person suddenly stops whatever he or she is doing, seems "absent" for a few seconds, and then resumes the previous activity. Some patients experience several petit mal seizures per day. Petit mal epilepsy is hereditary, occurs predominantly in children, and never begins after age 20.

A **tonic-clonic seizure** often begins with an "aura" of distorted perception and consciousness, progresses to unconsciousness and falling, and then to *tonic-clonic* contractions of the musculature. *Tonic-clonic seizures* refer to an initial increase in muscle tone associated with the loss of consciousness (the *tonic phase*), followed by a period of alternating muscle contraction and relaxation (the *clonic phase*) that may last from 2 to 5 minutes.

ATP and oxygen consumption by the brain increase markedly during a tonic-clonic seizure. The brain may consume ATP faster than aerobic respiration can produce it. A severe seizure can cause hypoxia, lactic acid accumulation, and cerebral acidosis, leading to the destruction of brain tissue. Death may ensue from cerebral necrosis or aspiration of fluid into the lungs.

Case Study 14 The Student with Meningitis

Heather, a 22-year-old college senior, spends a semester as an intern in Venezuela. Before coming home, she decides to see some of the sights and travels around the country for 3 weeks. Upon her return, she notices that she feels tired, her neck is stiff, her head aches, and she has a slight fever. She tells her roommate that she is going to lie down for a nap. Later that evening when Heather's mother calls, her roommate has difficulty waking her and notices that Heather seems to be having trouble walking and talking. After realizing that Heather has a fever, she takes Heather to the student health center. On the way, Heather becomes nauseous and vomits.

While Heather's roommate is checking her in at the health center, Heather becomes uncooperative and wants to leave. She doesn't calm down until one of the aides comes out. After this, Heather appears to relax and to "fall asleep with her eyes open." At this time, her roommate gives the nurse as complete a history as possible, including information about Heather's recent trip to Venezuela. In the examination room, the physician notes that Heather's neck is rigid and she has difficulty moving her legs. Her oral temperature is 104°F, and her heart rate is 130 beats/min. During the examination, Heather vomits again and begins to convulse. Suspecting meningitis or encephalitis, the physician immediately orders a spinal tap and blood sampling and starts intravenous antibiotics (ampicillin).

Results of the spinal tap and blood work are shown here.

CSF:

Leukocyte count = 20,000 cells/mL

Glucose = 18 mg/dL

Protein = 40 mg/dL

Diplococci present

Blood:

Glucose = 120 mg/dL

RBC count = $3.8 \times 10^6/\mu L$

WBC count = $18{,}000/\mu L$

(In both the CSF and the blood, the majority of leukocytes are neutrophils.)

Based on the results of these tests, Heather is diagnosed with bacterial meningitis. The intravenous antibiotic treatment is continued, but the antibiotic is changed to penicillin G, which is more specific for pneumococcus. A CT scan of the head is performed and shows no cerebral infarction and minimal inflammation. Heather is also given intravenous fluid and an antipyretic. When her parents arrive, they are told that Heather should recover fully.

Based on this case study and other information in this chapter, answer the following questions.

1. Why does the physician suspect either meningitis or encephalitis?
2. If the diagnosis is in doubt, why are antibiotics administered immediately?
3. How do the results obtained from the spinal tap and blood sample support the diagnosis? How do the results obtained from the spinal tap rule out viral encephalitis?
4. Other than meningitis or encephalitis, what conditions could account for some or all of Heather's signs and symptoms?
5. Why is fluid and electrolyte replacement necessary?
6. If inflammation of Heather's meninges had caused compression of the brain, how would the ventricles appear in the CT scan?
7. Polio compromises respiratory function in some patients, requiring the use of a mechanical ventilator. What happens at the cellular level to cause these breathing problems?
8. Recall that virtually all cases of polio in the United States now result from the polio vaccine itself. In light of this, would you choose to have

your child vaccinated against polio? Why or why not?

9. Which type of glial cell do you think is responsible for the scars that form in amyotrophic lateral sclerosis? Explain.

10. Al is brought into the emergency room after a motorcycle accident. At the scene, he was found unconscious, with numerous scrapes and bruises on the left side of his head and swelling above his right ear. When he regains consciousness 3 hours later, he complains of a headache and appears dazed. His speech is slurred, and his pupils are equal and reactive to light. Twenty-four hours later, all signs appear normal. Which of the following is the most likely diagnosis of Al's condition?
    a. mild concussion
    b. massive subdural hematoma
    c. severe concussion
    d. meningitis
    e. cerebral contusion

Selected Clinical Terms

apnea Temporary cessation of breathing.

concussion An injury to the brain or other soft tissue resulting from a violent blow or shaking.

contusion An injury to the brain or other tissue that causes subsurface hemorrhaging (bruising).

febrile Accompanied by fever.

hematoma A mass of clotted blood in the tissues; a bruise.

hemiparesis Muscular weakness or partial paralysis on one side of the body.

hemiplegia Complete paralysis of muscles on one side of the body, usually indicating a cerebral lesion.

hypotension Abnormally low blood pressure.

15 The Peripheral Nervous System and Reflexes

Objectives

In this chapter we will study

- the diagnostic value of testing somatic reflexes;
- the neurological basis of low back pain; and
- the epidemiology, pathogenesis, treatment, and prevention of rabies.

Reflex Testing

The peripheral nervous system (PNS) is composed of cranial nerves, spinal nerves, the distal branches of these nerves, and ganglia. Any disorder of nerve function is called a **neuropathy.** The spinal cord and spinal nerves are tested by assessing various somatic reflexes. Testing a reflex helps a clinician evaluate not only the individual components of a reflex arc (receptors, neurons, and muscles) but also the overall state of the nervous system. Abnormalities of reflex function, coupled with other information gathered during the physical examination, provide valuable clues to diagnosis.

One advantage of reflex testing is that it is easy to do; it requires only simple tools and good powers of observation. Although just a few reflexes are routinely tested, many others can be tested if necessary. Reflexes are usually graded on a scale of 0 to 4+:

0 **Areflexia,** absence of response

1+ **Hyporeflexia,** a somewhat diminished response

2+ An average or normal response

3+ **Hyperreflexia,** a stronger than normal response, possibly indicating disease

4+ Intense hyperreflexia with sustained clonus, indicating disease

Areflexia or hyporeflexia typically indicates a **segmental lesion** of the spinal cord segment or nerve root that innervates the muscle. Hyperreflexia typically indicates a **suprasegmental lesion** of higher levels of the CNS that normally inhibit the reflex.

The following discussion describes a few of the deep and superficial reflexes most commonly tested in adults and some of the reflexes tested in infants.

Deep Reflexes

The testing of deep reflexes usually involves striking the skin with a **reflex hammer** to stretch specific tendons and stimulate the tendon organs and muscle spindles. Clinicians commonly test the biceps, knee, and ankle reflexes and may test for clonus.

Biceps Reflex The patient lies supine with the elbow flexed about 30°. The clinician presses on the cubital fossa to stretch the biceps tendon and strikes his or her own fingers with the reflex hammer. In a normal response, the biceps should contract slightly, but not enough to flex the elbow. If there is a lesion of the musculocutaneous nerve or segment C6 of the spinal cord, the biceps does not contract but the finger flexors may contract slightly. If there is a suprasegmental lesion, the biceps may contract more forcefully than normal and the brachioradialis or finger flexors may contract.

Knee Reflex The patient either sits or lies supine with the knee flexed 90°. The examiner strikes the patellar ligament with the reflex hammer. Normal responses range from a slight twitch of the quadriceps femoris muscle to extension of the knee, lifting the leg. The absence of a response indicates a disease of lumbar nerve roots L3 and L4 or the femoral nerve.

Ankle Reflex There are several ways to test this reflex. One is to have the patient kneel on the examining table with the foot extending beyond the end of the table. The examiner presses slightly against the foot to dorsiflex it, thus stretching the gastrocnemius muscle, and then strikes the calcaneal tendon with the reflex hammer. The gastrocnemius should contract and plantar flex the foot. Other positions and methods can be tried if this one fails, but if no response is obtained by any method, a disease of the first sacral nerve root or the tibial nerve is indicated.

Clonus Clonus was described in the preceding chapter in connection with epilepsy (clonic seizures), but it can also be elicited in normal persons by the proper test. The patient should lie supine with the hip and knee flexed at 30° to 45° angles. The examiner then produces a sudden and sustained contraction of the gastrocnemius and soleus muscles by passively dorsiflexing the foot. In normal people, the calf muscles contract, relax, and contract again for about two or three beats. This occurs because the contraction of one muscle stimulates the stretch receptors in the antagonistic muscle. When the antagonist contracts, it stimulates the stretch receptors in the original muscle and triggers a reflex contraction. In people with suprasegmental lesions, the clonus continues for as long as the examiner dorsiflexes the patient's foot.

Superficial Reflexes

Superficial reflexes are tested by stimulating the skin. Following are three examples of these tests.

Abdominal Reflex The patient must be supine and relaxed. The examiner strokes the skin of the abdomen with a pointed object such as a pencil or the handle of a reflex hammer, moving from the lateral margins of the abdomen toward the midsagittal plane along a given dermatome. Normally, the underlying muscle contracts and pulls the umbilicus toward the stimulus. An absence of response in a given dermatome may indicate lesions to spinal nerves or roots T7 to T11. The response is often absent, however, in elderly patients and people with lax abdominal muscles.

Cremasteric Reflex In males, stroking the inner, upper aspect of the thigh with a pin or pencil point causes the ipsilateral testicle (but not the scrotum) to rise, owing to contraction of the *cremaster muscle*. Lesions in spinal cord segments or nerve roots L1 to L2 or in the corticospinal tracts abolish this reflex.

Plantar Reflex To test this reflex, the patient must be supine with the lower limbs extended. The examiner strokes the sole of the foot firmly with the handle point of the reflex hammer, progressing from the heel toward the toes. Normal subjects show a *flexor plantar response* in which they quickly flex the hip and knee, dorsiflex the ankle, and adduct and plantar flex (curl) the toes. An abnormal *extensor plantar (Babinski) response* is a reliable, early warning sign of corticospinal tract disease; the patient *extends* and *dorsiflexes* the great toe and *abducts* (fans) the other toes. The extensor plantar response also sometimes occurs in persons unconscious from drug or alcohol intoxication.

Reflexes of Infants

Because the nervous system is not completely developed at birth, neurological examination of infants differs somewhat from the techniques used for adults. Normally, a neonatal examination is performed between 36 and 60 hours after birth. In addition to reflexes, the infant's motor pattern and body posture are observed. A normal infant has flexed limbs, and its head may be turned to one side. The lower limbs may be moving or kicking, and the infant is expected to become more active and to begin crying during the examination. On the other hand, certain responses are considered abnormal. For instance, an infant extending its limbs may have suffered intracranial hemorrhage. Asymmetric behavior of the upper limbs suggests brachial plexus palsy. Lack of increased activity during the examination suggests anoxia or intracerebral hemorrhage.

Infants are tested for the same reflexes as adults as well as some additional ones described here.

The **trunk incurvation (Galant) reflex** is tested by stroking the back from the shoulder to the buttocks or vice versa, about 1 cm from the midline. This stimulus should elicit contraction of the ipsilateral back muscles, causing the infant's shoulders and pelvis to curve toward the stimulus while the trunk curves away. This response normally disappears at 2 months of age. Its earlier absence may indicate a transverse spinal cord lesion.

The **grasp reflex** is evaluated by determining the infant's ability to forcefully grasp the examiner's hand when the ulnar palmar surface is stimulated. This reflex normally disappears at 3 to 4 months of age. Persistence of the reflex beyond 4 months may indicate cerebral dysfunction.

The **rooting reflex** is a response to tactile stimulation of the lips. When the corner of the baby's mouth is stroked, the baby opens its mouth and turns its head toward that side. When the midline of the upper lip is stroked, the baby extends its head, and when the midline of the lower lip is stroked, the jaw drops. This reflex disappears at 3 to 4 months of age, although sleeping infants exhibit it at slightly older ages. Absence of this reflex before 3 to 4 months indicates severe CNS disease.

The **startle (Moro) reflex** is a response to a sudden stimulus such as a jolt, a loud noise, or being dropped a short distance (supporting the baby in a supine position and suddenly lowering it about 2 feet). The normal response is for the infant to extend and abduct all four limbs and extend and fan the digits, then flex and adduct the limbs. Neurologic disease is suspected if this reflex persists beyond 4 months and is almost certain if it persists beyond 6 months. An asymmetric response may indicate hemiparesis, brachial plexus injury, or fracture of the clavicle or humerus. The absence of a startle reflex may indicate *kernicterus*—damage to the basal nuclei or other areas of the CNS by accumulated bilirubin, a hemoglobin breakdown product seen in hemolytic disease of the newborn and some other conditions.

Diseases Affecting the PNS

Low Back Pain

Pain along the course of a nerve is called **neuralgia.** Because of the phenomenon of referred pain, the origin of pain can be difficult to identify—that is, pain may seem to come from the muscles or skin of a region when it actually originates in the abdominal or pelvic viscera. Sometimes pain originates in the ligaments of the spine, which are often damaged and which have abundant pain receptors.

Abnormal pressure on the peripheral nerves of the lumbar to sacral regions can cause *low back pain.* As noted in chapter 9 of this manual, low back pain may be idiopathic or may result from various vertebral disorders that create pressure on the spinal nerves. Two common causes are *degenerative disc disease* and *herniated (ruptured* or *slipped) disc*. For example, the pressure on the spinal nerves from a herniated disc causes pain that radiates along the sciatic nerve into the gluteal region and as far as the ankle. Even the strain of coughing or sneezing can trigger the pain. Pressure on a spinal nerve can also cause muscular weakness of the foot, paresthesia, and reduced sensations of touch, temperature, and pain.

The cause of low back pain, if it can be identified at all, may be determined through CT or MRI scans, specialized X-ray techniques, electromyography, and nerve conduction tests. Treatment typically involves bed rest, heat or ice, analgesics and anti-inflammatory drugs, and sometimes traction. Spinal surgery is of limited value in treating low back pain, but may be indicated if other treatments fail, if there is evidence of severe nerve compression, or if there is a loss of deep tendon reflexes or bladder or bowel control.

Rabies (Hydrophobia)

Rabies is an acute viral encephalitis that is usually transmitted by the saliva of infected mammals. It can affect any species of mammal. Worldwide, dogs are the most common source of human rabies because they interact so extensively with wildlife and humans; cats present a similar danger. Canine and feline rabies are now quite rare in the United States because of vaccination of pets—yet cats have twice the incidence of rabies that dogs do because people are less conscientious about keeping their cats vaccinated. Most human cases in the United States result from the bites of bats and other wild mammals. Many people who are bitten by bats are unaware of it, because most bats have tiny teeth and produce barely perceptible scratches on the skin. Raccoons, skunks, and foxes are common carriers of rabies but are rarely the source of human rabies infections. Although raccoons are the most common carriers of rabies in the United States, no one has ever gotten rabies from a raccoon bite, perhaps because people bitten by raccoons are well aware of it and are more likely to cleanse the wound and seek medical attention. Horses, cattle, and other livestock become infected from the bites of foxes and skunks. They can transmit the virus to humans when their owners examine the animals' mouths. Infected horses and cattle also sometimes pursue and bite humans. Rabbits and small rodents such as mice, chipmunks, and squirrels seldom transmit rabies to humans. In a few cases, humans have contracted rabies by inhaling the dust in bat-infested caves. A few people have died after receiving corneal transplants from donors with rabies.

Rabid animals exhibit either *dumb rabies* or *furious rabies.* In **dumb (paralytic) rabies,** animals show signs of paralysis, especially of the pharyngeal and masseter muscles. They salivate profusely and cannot swallow. Dogs with dumb rabies often wander about with the jaw open. In **furious rabies,** the "mad dog" form of the disease, animals are agitated, viciously aggressive, and show no fear of humans or their other natural enemies. (The word *rabies* is from the Latin *rabio,* "to rage.") Rabies should be suspected in wild mammals that seem unafraid of approaching humans and in normally nocturnal mammals (such as skunks, raccoons, bats, and foxes) that become uncharacteristically active during the day.

Pathogenesis Following a bite by a rabid animal, the saliva-borne virus replicates in the skeletal muscle at the site of the bite, then invades the nerves that supply either the muscle spindles or extrafusal fibers of the muscle. The virus travels about 3 mm/hr up the peripheral nerve fibers until it reaches the CNS. There, it multiplies in the gray matter and then travels down the autonomic nerve fibers to such sites as the adrenal medulla, kidneys, lungs, liver, heart, skin, and salivary glands. Infection of the salivary glands facilitates transmission to a new host.

The first stage of the infection is an **incubation period** in which the virus travels and replicates in the body but produces no noticeable symptoms. This can last from 10 days to more than a year in humans; it is usually 30 to 50 days long and on the short end of this range in people with bites on the face or trunk, people with multiple bites, and children. This is followed by a **prodromal phase** about 1 to 4 days long in which a person experiences vague, nonspecific signs and symptoms such as malaise, fever, fatigue, headache, nausea, and sore throat. The prodromal symptom most indicative of rabies is *paresthesia* (see chapter 13 of this manual), abnormal sensations that may result from the infection of dorsal root ganglia by the virus. Paresthesia is experienced by 50% to 80% of patients.

The third stage is *encephalitis*. As it sets in, the victim experiences confusion and deranged thoughts and may exhibit agitation, combativeness, muscle spasms, seizures, paralysis, and extreme sensitivity to stimuli such as light and sound. This phase is soon followed by brainstem and cranial nerve involvement characterized by double vision, visual hallucinations, facial palsy, excessive salivation, and difficulty swallowing. In about 50% of cases, attempts to swallow produce intensely painful muscle spasms of the pharynx and larynx. Infected humans and animals thus develop an aversion to water even though they are intensely thirsty. The alternative name of rabies, *hydrophobia,* refers to this seeming "fear of water." Not long after this stage, the patient lapses into a coma and dies of respiratory arrest.

By the time the first symptoms are felt, the virus has already invaded the CNS, and death is essentially 100% certain. Therefore, early diagnosis is of no help to the patient. However, the sooner a diagnosis of rabies is confirmed, the sooner preventive treatment may be initiated for others who have come in contact with the patient.

Diagnosis Rabies is typically diagnosed from the patient history (having suffered an animal bite) combined with the rapid neurological degeneration of the patient's condition. The signs and symptoms of rabies are ambiguous, however, and it is difficult to distinguish rabies from several other neurological diseases that are much more common and probable. A fluorescent antibody test for the rabies antigen confirms the diagnosis of rabies but comes too late to save the patient.

Because of the seriousness of rabies, when an apparently healthy dog or cat bites a person, the animal should be observed for a period of 10 days. If the animal shows no signs of rabies, treatment is not initiated. If the animal does exhibit signs of rabies, it is killed and its head is sent to a diagnostic laboratory. The brain tissue is histologically examined for diagnostic **Negri bodies,** small spherical masses of rabies antigen and viruses. The absence of Negri bodies does not necessarily mean an absence of rabies, however. If none are found, the brain tissue is further examined with the fluorescent antibody test.

Treatment An animal bite should be allowed to bleed freely and then promptly scrubbed with soap and flushed with water. Benzalkonium chloride, a germicide present in medicinal soaps, deactivates the rabies virus. Antibiotics and tetanus immune globulin should be given. The wound should not be cauterized or sutured.

If the animal that inflicted the bite is confirmed to have rabies, or if it escaped but is suspected to have had rabies, the patient is treated immediately with *rabies immune globulin (RIG)* and an antirabies *human diploid cell vaccine (HDVC)*. RIG is given as soon as possible after the exposure, and only once. It provides **passive immunity** to the rabies virus that serves until the patient builds up his or her own **active immunity** in response to the HDVC. The first dose of HDVC is given with the RIG on "day 0," followed by additional HDVC doses at 3, 7, 14, and 28 days (and sometimes 21 and 90 days).

Prevention People at high risk of contact with the rabies virus—animal handlers, laboratory personnel, veterinarians, and cave explorers—should be vaccinated with HDVC, get periodic evaluations of their antirabies antibody levels, and receive booster doses when their antibody levels fall below a critical value. The risk of exposure to rabies is greatly reduced by vaccinating dogs and cats, controlling stray animals, and strictly avoiding contact with wild mammals. It is important to bear in mind that wild animals can be infected and can transmit the virus even in their prodromal period, when they show no obvious signs of disease.

Case Study 15 Getting Close to Nature—Too Close

Michael invites his college roommate, Steven, to spend spring break with him at his family's hunting cabin in the hills of Pennsylvania. The cabin is not well kept; it is drafty and has a broken window. One night, Steven is awakened by something crawling on his right shoulder. It feels furry as he brushes it onto the floor. He guesses that it was a mouse, but doesn't get up to inspect it. In the morning, however, Michael finds a bat clinging to one of the log walls. It makes strange noises and flies about the room, but it appears unable to fly very well and repeatedly lands on the floor. Michael forces the bat out the cabin door with a broom. It flies several feet away into the woods and lands on the ground.

Back at college a month later, Steven wakes up one morning with a headache and low back pain. He attends a morning class, but as the day progresses, his neck and shoulder begin to ache, his arm hurts and feels weak, and he begins to get a sore throat. He doesn't feel like eating, so instead of going to lunch, he goes back to the dormitory, takes two Tylenol, and lies down for a nap. When Michael comes in, he wakes up and tries to speak, but his voice is very hoarse. Michael notes that Steven seems a little confused and disoriented, and takes him to the campus infirmary.

At the infirmary, Steven is found to have a slightly elevated temperature, and his sore throat feels worse. He is given more Tylenol, an antibiotic, and an anesthetic throat spray and told to check back if his condition changes. The next day, Steven has a high fever (104°F) and chills, combined with nausea and vomiting. The pain in his neck and shoulder is intense, and his speech is slurred. Now increasingly concerned for his friend, Michael drives Steven to the emergency room of the county hospital.

At the hospital, a laryngeal examination shows paralysis of the vocal cords on the left side. Steven is now hypersalivating and cannot swallow his saliva without intense pain. He is admitted to the hospital for observation. A few hours later, he begins to exhibit clonus of the right arm, progressing to generalized clonus of the trunk and other limbs. He is given anticonvulsive medication and undergoes several tests. The results of an electroencephalogram reveal no evidence of epileptiform seizure, and the results of CT and MRI scans appear normal.

Over the following day, Steven requires continual oropharyngeal suction to remove excess saliva and is placed on I.V. fluid to prevent dehydration. The doctors suspect tetanus because Steven has had a work-study job in the biology department greenhouse. They give him tetanus immune globulin and continue antibiotic treatment. They also consider herpes simplex encephalitis and spongiform encephalopathy to be possibilities. Steven now begins to have double vision and to see odd flashes of colored light and other hallucinations.

Samples of blood serum, saliva, CSF, and a nuchal skin biopsy are sent to the Centers for Disease Control for diagnostic examination. While doctors await the results, Steven goes into respiratory arrest and is put on a mechanical ventilator. The results of the serum and CSF analyses come back negative, but Steven's saliva and skin biopsy test positive for rabies. Eight days after the onset of the symptoms, Steven can no longer breathe on his own, his pupils are unreactive, and he shows no corneal reflex. At 12 days, he shows no cranial nerve reflexes at all, and lapses into a coma. On day 14, Steven's parents consent to withdraw respiratory support, and Steven dies.

Upon autopsy, no brain tumors are found, and there is no gross evidence of cerebral necrosis or hemorrhage. Samples of brain tissue are sent to the CDC, where histological examination shows microscopic necrosis and the presence of Negri bodies, especially in the cerebellum and hippocampus. A fluorescent antibody test confirms the presence of rabies antigen. An RNA sequence analysis shows that Steven was infected with a variety of the rabies virus associated with two species of bats, the silver-haired bat and the eastern pipistrelle, both known to be common carriers of rabies.

Because of the rabies diagnosis, the hospital begins post-exposure prophylaxis (PEP) of the people who were or may have been exposed to Steven's saliva, including Steven's roommate, his parents and sister, his girlfriend, and 42 nurses, doctors, orderlies, and other staff of the campus infirmary and the county hospital. All 47 people receive an initial dose of rabies immune globulin (RIG) and five doses of rabies vaccine (HDVC) over the next 28 days.

Based on this case study and other information in this chapter, answer the following questions.

1. How could this tragedy have been prevented? What should Steven have done, and when?
2. Name the stage of the disease that Steven is in on the day he first goes to the campus infirmary. Could his death have been prevented if the disease had been correctly diagnosed on that day? Explain.
3. From what you know of the pathogenesis of rabies, explain why the disease progresses less rapidly in a person bitten on the leg than in a person bitten on the neck or shoulder.
4. Why are so many different tissue and fluid samples from Steven sent to the CDC? From what part of the body is the skin biopsy taken? (Use a common term, as if you were explaining this to someone with no background in human anatomy.)
5. What does Steven's work-study job have to do with the doctors' tentative diagnosis of tetanus?
6. Steven's parents request that his organs and tissues be used for transplant to other patients. Do you think the hospital would honor his parents' wish? Why or why not?
7. What sign do rabies and grand mal epilepsy have in common? In what sign do they differ?
8. What is the adaptive significance (survival value) of the rooting reflex?
9. How does paresthesia differ from reduced sensitivity to touch, temperature, or pain?
10. Low back pain could result from any of the following except
    a. rabies.
    b. driving earth-moving machinery for a living.
    c. competitive weight-lifting.
    d. hyporeflexia.
    e. osteoporosis.

Selected Clinical Terms

active immunity Immunity that results from one's own production of antibodies or immune cells against a pathogen.

clonus The pulsating, repetitive contraction and relaxation of a muscle or muscle group. A few pulsations are normal when clonus is properly tested, but persistent clonus can indicate certain seizure disorders such as epilepsy and other neuropathies such as rabies.

incubation period A period in the course of an infectious disease between the time the pathogen enters the body and the appearance of the first signs or symptoms of disease; also called *latent period.*

Negri body A microscopic round mass of rabies virus and rabies antigen found in the cytoplasm of neurons (especially in the cerebellum and hippocampus) of infected people and animals; serves as a histopathological confirmation of rabies.

neuralgia Nerve pain; especially sharp stabbing or throbbing pain along the course of a nerve.

neuropathy Any disorder of nerve function—for example, sciatica, rabies, or shingles.

passive immunity Immunity conferred by antibodies or immune cells acquired from another donor, usually through injection (vaccination).

prodromal phase A period in the course of a disease when the first symptoms are felt and a person has a premonition of oncoming illness.

segmental lesion An injury to a segment of the spinal cord through which passes a particular reflex arc that is being tested; tends to produce hyporeflexia or areflexia.

suprasegmental lesion An injury to a level of the CNS higher than the segment of the spinal cord involved in a particular reflex arc that is being tested; tends to produce hyperreflexia because of a loss of normal inhibitory influence.

16 Sense Organs

Objectives

In this chapter we will study

- methods used to test the function of the general senses;
- the classification, treatment, and management of pain;
- methods used to test the special senses of hearing and vision;
- some treatments for deafness; and
- conjunctivitis and retinitis pigmentosa.

Investigating Disorders of the Senses

The *general senses* (also called the somesthetic senses) function through sensors that are widely distributed in the body rather than limited to specific localities. These receptors in the skin, muscles, tendons, joint capsules, and viscera detect touch, pressure, strength, heat, cold, and pain. The *special senses* are vision, hearing, equilibrium, taste, and smell. Their receptors are located only in the head, where they are innervated by the cranial nerves.

Investigating sensory disorders requires an understanding of the relationship between the sensory systems and the central nervous system. By examining the interactions among the sensory receptors, sensory nerves, ascending pathways, brain nuclei, and regions of the cerebral cortex, a clinician should be able to determine the location of a disorder and suggest a course of action.

The General Senses

Testing the General Senses

The general senses are typically tested during a neurological examination. Some of the methods of testing them are briefly described here:

- **Light touch** The sense of light touch can be tested by touching various regions of the body with a few fibers teased from a cotton ball. It is important not to drag the cotton wisp across the skin because this induces the sensation of tickle rather than touch. The patient should be able to feel at least 18 of 20 touches on any area of skin.
- **Superficial pain** The examiner touches the skin in several regions with both the point and the head of a common pin, asking the patient, "Is this sharp or dull? Does this feel the same as the previous touch?" and so forth.
- **Temperature** A patient's ability to detect heat and cold is tested by touching the skin with dry test tubes filled with hot and cold tap water. Thermoreceptors respond slowly, so the test tube must be held against the skin for at least 2 seconds at each spot tested. The examiner asks the patient whether the stimulus feels "as hot" or "as cold" in an area of suspected dysfunction as it feels in a normal area.
- **Vibration** To test the vibration sense, the clinician touches the skin with the handle of a vibrating tuning fork and asks the patient to report whether he or she feels the vibration and when it stops. A little teaching is needed before the test to ensure that the patient can discern the difference between the vibration and the pressure of the tuning fork and between a vibrating and a nonvibrating tuning fork. (The latter is tested by placing the tuning fork on the jaw or sternum.) Then the clinician can test other areas of the body. The vibration of the tuning fork gets weaker with time, and the patient is typically asked whether he or she senses the vibration initially and when it stops.
- **Proprioception** To test the patient's knowledge of the movement of the joints, the clinician moves the terminal phalanges of the patient's hands and feet up and down while the patient's eyes are closed and asks the patient to identify which way the joint is being moved. Some neuropathies affect the distal parts of a nerve more than its more proximal regions. Therefore, if a patient fails to feel the movement of a distal joint, more proximal

joints are tested to assess the extent of the nerve damage.

- **Equilibrium (balance)** Lesions of the vestibular apparatus of the inner ear, the vestibulocochlear nerve, the cerebellum, and other parts of the nervous system can affect a person's ability to maintain balance or execute smooth movements. Since people tend to visually compensate for and maintain their balance by viewing their surroundings, it is necessary to eliminate the visual component when testing for sensory, nerve, or CNS lesions. One way of doing this is the **Romberg test.** The patient is asked to stand with his or her feet as close together as possible, as long as he or she feels stable and comfortable, and then to close the eyes. Patients with a defective sense of balance lose their balance and do not merely teeter a little (which is normal), but fall to one side when they close their eyes.

When testing the skin senses, it is important to vary the timing and placement of stimuli so that the patient cannot predict when or where the next stimulus will occur. Most patients fail to feel some stimuli and imagine others when no stimulus has been applied. Skillful questioning is also necessary to elicit the most helpful responses from the patient.

The clinician looks for unusually high or low sensitivity, differences in sensitivity between one side of the body and the other, and differences from one dermatome to another. The distribution of any abnormalities helps distinguish lesions of the brain, spinal cord, and peripheral nerves.

Diagnosing and Managing Pain

Treatment for pain depends first on its proper diagnosis and classification. There are several ways to classify pain. One system is *somatogenic* versus *psychogenic* pain. **Somatogenic pain** has a physical cause, such as a twisted ankle or laceration, while **psychogenic pain** has a psychological cause. It is important to remember that psychogenic pain is no less "real" to the patient and requires appropriate treatment. In many cases, pain has a mixture of somatogenic and psychogenic causes—that is, we are often prone to think that an injury hurts more than it "really does," and the level of pain we experience is often a matter of attitude.

Pain is also classified as either *acute* or *chronic* according to its duration. **Acute pain** is usually short-lived and serves as a protective mechanism, alerting the body to damage. It has a sudden onset and is relieved fairly readily by taking analgesics or removing the stimulus. **Chronic pain** is usually of extended duration (longer than 3 months). Its cause is often unknown, and it is therefore more difficult to treat.

The body responds differently to acute and chronic pain. Acute pain stimulates autonomic responses such as increased heart and respiratory rates, blood pressure, and blood glucose concentration, dilation of the pupils, and decreased gastrointestinal secretion and motility. In effect, the body is responding to actual or potential tissue injury in such a way as to restore homeostasis. Chronic pain produces fewer obvious physiological effects because, over the long run, the body adapts to it and maintains homeostasis. Chronic pain does, however, produce significant psychological and behavioral changes, such as depression, insomnia, anorexia, preoccupation with the pain, and feelings of helplessness. Patients experiencing chronic pain may exhibit adaptive behaviors in an attempt to alleviate the pain or to maintain a "normal" appearance to others. Some individuals tend to not report the pain or even to deny it. Neglect of the underlying cause can then make the pain grow more severe.

The point at which an individual perceives a stimulus as painful is the **pain threshold**. In general, there is not a large degree of variation in the pain threshold among individuals or within the same person over time. The subjective intensity of pain, or the length of time an individual can endure it before responding noticeably, is **pain tolerance.** A great degree of variation in pain tolerance exists among individuals and within the same individual over time. Several factors affect a person's pain tolerance, including physical and mental state, cultural background, gender role, age, and expectations. In general, pain tolerance is decreased by fatigue, anger, apprehension, boredom, and repeated exposure to pain. It is increased by medication, alcohol, hypnosis, warmth, distracting activities, and strong beliefs or faith.

Infants cannot articulate their pain as a child or adult can, but they do show stereotyped responses that indicate they are in pain: brows drawn together and lowered, vertical furrows in the forehead, bulges between the brows, tightly closed eyes, raised cheeks, a bulging, broadened nose, quivering chin, and an open, squarish mouth. The blood pressure, heart rate, and respiratory rate are elevated, and the infant may sweat and appear flushed.

In evaluating patients who complain of either acute or chronic pain, the clinician first obtains a detailed history of the pain—its severity, location, duration, course, timing, factors causing it to worsen or ease, drug use (either for pain relief or other reasons), and any other associated symptoms (for example, psychological state). The patient's level of function is also determined. A physical examination is completed to identify whether any underlying causes exist and to determine the need for laboratory tests. Based on these results, the clinician decides on a course of action to manage the pain.

Pain management varies considerably from patient to patient and may include analgesics, physical therapy, rest, or surgery. For some patients, psychogenic pain can be alleviated with placebos, hypnosis, or biofeedback.

The Special Senses

Hearing

Deafness is any partial or complete loss of hearing. There are many types of deafness, but they all fall into the two categories of *conductive deafness* and *sensorineural deafness*. **Conductive deafness** results from the inability to transfer vibrations through the auditory canal and middle-ear ossicles to the inner ear. Its causes range from impacted earwax to **otosclerosis,** a pathological calcification of the middle-ear ossicles that "freezes" them so that they cannot vibrate freely. **Sensorineural deafness** (formerly called *nerve deafness*) results from defects in the cochlea, vestibulocochlear nerve, or CNS. The most common cause is cochlear damage resulting from exposure to loud noise; this condition is frequently seen in musicians (especially those who play in bands with amplified music and in symphony orchestras), people who listen to excessively loud recorded music, machinery operators, and people who use firearms without ear protection.

The two forms of deafness can be distinguished from each other by a variety of hearing tests.

Hearing Tests As a preliminary assessment of a patient's hearing, an examiner may ask if the patient can hear on the telephone through both ears. Another method is conducted in a quiet examination room by standing 6 feet to one side of the patient and asking him or her to put a finger in the opposite ear and try to repeat a series of numbers that the examiner whispers. A normal person should be able to repeat at least 9 out of 10 numbers.

If a hearing deficiency is reported or observed, it is important to determine whether the cause is conductive or sensorineural. This can be done with the **Rinne test.** The clinician strikes a tuning fork and holds the stem against the mastoid process behind the patient's ear. The patient normally hears a vibration because of **bone conduction**—that is, the tuning fork vibrates the skull, and this stimulates the cochlea. When the patient no longer hears it, the examiner immediately holds the tuning fork (still faintly vibrating) next to the patient's auditory meatus. People with normal hearing or partial sensorineural deafness again hear the hum because air conduction is more efficient than bone conduction. But in people with conductive deafness, the airborne sound disappears more quickly than the bone-conducted sound, so by the time the tuning fork is moved from the mastoid process to the auditory meatus, the vibration is too faint for the patient to hear.

Audiometry is a more precise measure of the type and degree of hearing loss. The patient listens with headphones to an electronic device called an **audiometer** that plays pure tones of specific frequencies and volumes. The audiometrist measures the volume (loudness) required for the patient to hear pure tones ranging from about 250 to 8,000 Hz and charts an **audiogram** showing the degree of hearing loss versus the sound frequency.

Auditory brainstem responses can be used to test hearing in patients who cannot or will not report whether they hear a sound, such as comatose patients, infants, and people feigning deafness. An electroencephalogram is recorded while the patient is given an auditory stimulus. If the ear, vestibulocochlear nerve, brainstem, and auditory cortex are functioning properly, the EEG shows characteristic responses to sounds.

Correcting Deafness Hearing aids can benefit some patients with conductive or sensorineural deafness, depending on the degree of hearing loss and the frequency range in which the loss exists. Several types of hearing aids are available, with different benefits and drawbacks for different kinds of patients. Patients can also benefit from training in **speech reading** (lip reading) provided by speech pathologists associated with audiologists. Some people with hearing loss too profound to be helped by hearing aids can benefit from a **cochlear implant.** This is an electronic device implanted beneath the skin behind the ear, with electrodes that lead to the cochlear nerve. The implant detects sound and randomly stimulates cochlear nerve fibers. The resulting sensation does not resemble normal hearing, but patients can learn to associate the frequency responses they hear with the origin and relevance of a sound. It aids in distinguishing the rhythm of speech, when words begin and end, and other speech qualities that, combined with speech reading, can make conversation intelligible. A cochlear implant also helps deaf persons hear important environmental noises such as alarms and car horns.

Vision

Our ability to see relies on the structures of the eyes as well as various components of the central, somatic, and autonomic nervous systems. Thus, a vision examination can provide information about not only the eyes but also the thalamus, visual cortex, brainstem, cerebellum, autonomic nervous system, and cranial nerves.

Eye Examinations Because the eye, its accessory organs, and the visual process are so complex, a comprehensive eye examination involves more tests and observations than does any other sensory examination. Following are the major tests and observations and some of the ways they can aid diagnosis.

- **External anatomy of the ocular region** The eyebrows, eyelashes, underlying skin, and eyelids are examined for the quantity and distribution of hair, scaliness of the skin, edema of the eyelids, adequacy of eye closure, and other properties. Certain abnormalities can indicate hypothyroidism, seborrheic dermatitis, and other pathologies. Drooping eyelids *(ptosis)* may indicate an oculomotor nerve lesion or myasthenia gravis. The lacrimal apparatus and degree of eye moisture are also observed. Abnormalities can indicate conjunctival inflammation or obstruction of the nasolacrimal duct. The conjunctiva and sclera are examined by having the patient look up while the examiner depresses the lower eyelids looking for signs of inflammation (redness), jaundice (yellowness), and other abnormalities. The general position and alignment of the eyes are noted. Bulging eyes *(exophthalmos)* are a sign of hyperthyroidism (Graves disease).
- **Cornea, lens, iris, and pupil** The cornea and lens are examined for clarity. Grayish-white scars in the cornea are signs of old eye injuries and inflammation. Clouding of the lens is known as *cataracts,* a condition frequently seen in elderly people. The iris should be flat and exhibit clearly defined markings. A rounded iris, which casts a shadow when illuminated from one side, can be a warning sign of glaucoma. Slight inequality of the pupils is common, but inequalities of more than 0.5 mm diameter and defective pupillary responses to light may indicate glaucoma, oculomotor nerve dysfunction, or other disorders.
- **Intraocular pressure** Glaucoma is caused by a rise in intraocular pressure. Even people who do not need corrective lenses or a change in lens prescription should have periodic eye exams that include measurement of intraocular pressure. This is done with an instrument called a **tonometer.** The tonometer shines a small beam of light on the eye and tracks its movement as (in one design) a puff of air is blown on the eye. The greater the intraocular pressure, the more the eye resists indentation by the puff of air, and the less deflection of the light beam is recorded.
- **Eye movements** The examiner asks the patient to visually follow a moving object, such as a finger or pencil, through an H-shaped pattern—to the extreme right, up and down on the right, to the extreme left, and up and down on the left—and then observes the convergence of the eyes as the object is moved toward the patient's nose. This tests the extraocular muscles and the nerves that control eye movements—the oculomotor, trochlear, and abducens. A rhythmic oscillation or flickering of one or both eyes, called *nystagmus,* can indicate lesions of the vestibular apparatus or brainstem or result from the abuse of alcohol and other drugs. *Strabismus,* in which one eye looks in a different direction than the other, indicates weakness of one or more extraocular muscles.

- **Visual acuity** Visual acuity—the ability to see clear images—is usually tested with a *Snellen eye chart* mounted on a wall. Normally, the patient should be 20 feet from the chart, cover one eye, and read the smallest possible line of print with the other eye. Visual acuity is expressed as a ratio of two numbers, such as 20/30. An acuity of 20/30 means that with that eye, the patient can read print from a distance of 20 feet that a normal person can read at 30 feet. Each eye has its own acuity, so a person's vision could be 20/30 in one eye and 20/40 in the other. In the United States, a person is considered legally blind if, with corrective lenses, he or she has a visual acuity of no better than 20/200 in either eye.

- **Visual field** The *visual field* is the area in space that a person can see without moving the head. To evaluate a patient's visual field, the examiner can hold his or her arms outstretched to each side of the patient's head, and slowly bring them forward while wiggling the fingers, following an arc as if tracing the surface of a glass bowl in front of the patient's face. The patient is told to indicate when he or she first sees the examiner's fingers in the peripheral vision. If a defect is noticed, the examiner uses more refined tests to identify the area and extent of the defect. Another U.S. definition of legal blindness is that neither eye can see a visual field of more than 20º. *Field defects* in vision include blindness in the right or left half of the visual field of one or both eyes *(homonymous hemianopsia),* blindness in the lateral half of the field in both eyes *(bitemporal hemianopsia)*, blindness in one-quarter of the visual field *(quadrantic defects),* and patches of blindness surrounded by areas of normal vision *(scotomas).* Field defects can be caused by lesions of the retina, optic nerve, optic chiasm, optic tract, and optic radiations of the brain. A lesion in each of these locations produces its own characteristic field defect.

- **Ophthalmoscopic examination** A comprehensive eye examination also includes a visual inspection of the interior of the eye, using an illuminating, magnifying instrument called an **ophthalmoscope**. The retina appears as a red-orange circle. Toward the medial (nasal) side is the *optic disc,* the head of the optic nerve. It is much paler than the rest of the retina, with a pink to yellow-orange color, and has an array of retinal arteries and veins converging on it. At the center of the retina is a darker red patch, the *macula,* with a yellow-white depression, the *fovea centralis,* which is our area of sharpest vision. The examiner inspects the size, color, and shape of the optic disc and the health of the retinal arteries and veins, looking for pathologies such as cloudiness in the lens and vitreous body, hemorrhages or lesions of the retina, and retinal wrinkling or detachment. Hypertension and diabetes mellitus are among several diseases that can be detected in part by ophthalmoscopic examination.

Eye Disorders

Conjunctivitis is inflammation of the conjunctiva resulting from allergies, viruses, and bacteria. It is characterized by hyperemia (abnormally increased blood flow) of the eye and eyelids, redness of the conjunctiva, burning pain, and discharge. The discharge is usually cultured to determine the causative agent. Conjunctivitis is readily spread by contact, so it is important that the patient follow hygienic practices such as not sharing towels with other people. Treatment most often involves removing any irritating agent, keeping the eyes free of discharge, and applying antibiotic eyedrops.

Retinitis pigmentosa (RP) is a bilateral degeneration of the retina. It is an untreatable hereditary disorder that results in the progressive loss of night vision and peripheral vision, thus progressively narrowing the central visual field. Ophthalmoscopic examination shows a waxy, pale optic disc, constriction of the peripheral retinal blood vessels, and black "spicules" of pigment clustered around the vessels. The patient experiences ringlike or arclike scotomas encircling the central vision. The disease often appears early in childhood. Autosomal recessive, autosomal dominant, and X-linked forms of RP are known.

Case Study 16 No More Loud Music

Roger, age 40, goes to his doctor for a regular physical examination. In the course of the interview, the doctor asks him if his hearing is okay. Roger says that his wife sometimes accuses him of being "half deaf" because he turns the stereo and television up louder than she likes it. He also says has been late for work a few times because he didn't hear the high-pitched beeping of his alarm clock, so he bought a clock radio and set it to wake him up to music instead. The doctor examines Roger's ears but sees no impacted cerumen or other abnormalities. He asks if Roger would like to be referred to an otolaryngologist, and Roger agrees that it would be a good idea to have his hearing checked more thoroughly.

The otolaryngologist tells Roger that his receptionist will make an appointment for him to see an audiologist, but says that he can do a few simple tests in the meantime. Roger tells him the same things he told his regular doctor, but adds that he often finds it difficult to make out a person's words—for example, when watching television or at the theater—making it hard for him to follow the plot of a movie. He says that he can still hear telephone conversations with both ears, however. Roger says that he has liked his music loud ever since college and often plays his car stereo loud and with heavy bass amplification. He also likes to go to drag races three or four times a year. The otolaryngologist asks if Roger has ever worked in a loud environment such as a factory, ever used firearms without ear protection, ever served in combat, and other questions—to all of which Roger answers no. To other questions, however, he answers yes: He had frequent middle-ear infections when he was a child, and he had mumps when he was 8 years old—conditions that can cause hearing loss.

The doctor performs a Rinne tuning fork test. With the tuning fork against his mastoid process, Roger hears the hum for 16 seconds in the left ear and 18 seconds in the right ear (bone conduction times). With the tuning fork then moved in front of the auditory meatus, Roger hears the hum for an additional 12 seconds with both ears (air conduction time). The doctor remarks that this suggests a sensorineural hearing loss, not a conductive hearing loss.

Roger sees the audiologist the following week. The audiometer plays pure tones at selected intensities and frequencies from 250 to 8,000 Hz. At frequencies from 300 to 3,000 Hz, the range of most conversation, Roger has a hearing loss in both ears ranging from 40 to 60 decibels (db). Specialized audiometric tests that distinguish bone conduction from air conduction confirm that Roger has sensorineural deafness rather than conductive deafness. The otolaryngologist tells Roger that his earlier middle-ear infections probably haven't affected his hearing. His mumps could have something to do with it, but more likely, the loss is a result of inner-ear damage from the loud music. He advises Roger that he must be more careful to protect his hearing, and that if his hearing loss is too troublesome, he may want to consider a hearing aid and training in speech reading so that he will be able to better understand movies, conversations, and so forth.

Based on this case study and other information in this chapter, answer the following questions.

1. Suppose Roger had been a hunter and regularly used a shotgun without using ear protection. Which type of hearing loss—conductive or sensorineural—would most likely result from this activity?
2. Why did the otolaryngologist not suggest cochlear implants for Roger?
3. Suppose that at 2,000 Hz, Roger has a 50-db hearing loss in his left ear. How many times louder must a 2,000-Hz tone be for Roger to hear it than it would for a person with perfect hearing to hear it?
4. Why does the otolaryngologist conclude that Roger's childhood middle-ear infections are not responsible for his hearing loss?
5. Suppose the doctor had held the tuning fork in front of Roger's auditory meatus until Roger could no longer hear it, and then held the stem of the tuning fork against Roger's mastoid process. How long do you think Roger would be able to hear it after it was moved?

6. If you were testing a drug for its analgesic effect, why would measuring the change in the patient's pain threshold be more relevant than measuring the change in the patient's pain tolerance?
7. Which of the following would *not* suggest conductive deafness?
   a. inability to hear your own words
   b. wax accumulation in the ears
   c. inability to hear softly spoken words
   d. inability to hear the bass in a musical piece
   e. inability to hear well after swimming
8. If conjunctivitis were left untreated, what part of the eye do you think would be affected next?
9. Do you think retinitis pigmentosa has an earlier and more serious effect on rod cells or on cone cells? Explain your answer.
10. Why would it be important not to let a patient watch your procedures if you were testing his or her general senses such as touch, pain, and proprioception?

Selected Clinical Terms

audiometer An electronic device that produces pure tones of selected frequencies and volumes for the purpose of testing the threshold of hearing over a range of 250 to 8,000 Hz.

conductive deafness Hearing loss resulting from any failure to transmit sound vibrations to the inner ear, for example due to an obstructed auditory canal or abnormal ossification of the joints between the auditory ossicles.

conjunctivitis Inflammation of the conjunctiva.

retinitis pigmentosa A progressive, hereditary disease in which the inner layers of the retina atrophy and become infiltrated with pigment.

Rinne test A hearing test that uses a vibrating tuning fork held against the skull and then in front of the auditory meatus to distinguish between conductive deafness and sensorineural deafness.

Romberg test A test of equilibrium in which a patient's balance is observed as he or she stands with the feet together and the eyes closed, so that visual cues cannot override any defects in the functionality of the vestibular system.

sensorineural deafness Hearing loss resulting from any defect in the cochlea, vestibulocochlear nerve, or auditory cortex of the brain.

tonometer An instrument used to measure pressure in the anterior chamber of the eye by measuring the resistance of the cornea to deflection by an external force such as a puff of air or a probe that taps the cornea.

17 The Endocrine System

Objectives

In this chapter we will study

- the general causes of and diagnostic approaches to endocrine disorders;
- four disorders of the pituitary gland—hypopituitarism, gigantism, acromegaly, and SIADH;
- disorders caused by a hyperactive or hypoactive thyroid gland; and
- disorders caused by a hyperactive or hypoactive adrenal cortex.

Overview of Endocrine Disorders

The internal coordination of the body depends primarily on the nervous and endocrine systems, so a dysfunction in either of these systems can cause widespread derangement of bodily function. Disruptions in endocrine function can involve hormone **hyposecretion** (deficiency) or **hypersecretion** (excess), abnormal secretion of hormones by **ectopic** sources (organs other than the endocrine glands), alterations in the rate of hormone degradation and clearance from the blood, and alteration of target organ responsiveness.

Diagnosing Endocrine Disorders

Endocrine disorders, like those of the nervous system, can be difficult to diagnose because of the wide range of interconnected signs and symptoms they produce. Some endocrine disorders exhibit distinct symptoms, such as the bronzing of the skin in Addison disease and the goiter in some types of hypothyroidism. In other disorders, however, the signs and symptoms can be vague and overlap with those of nonendocrine disorders.

As with all diagnoses, a patient history and physical examination provide the first clues to an endocrine disorder and suggest tests that should be conducted. Because hormones are carried in the blood and they or their metabolites are eventually excreted in the urine, hematology and urinalysis provide especially important information in diagnosing endocrine disorders. Other tests are available to test the function of specific endocrine glands. For example, diabetes mellitus is diagnosed partly from a *glucose tolerance test*. The patient ingests a high-sugar solution, and then the level of blood glucose is monitored over time to see how quickly it is absorbed from the blood. A prolonged, high glucose level indicates either a deficiency of insulin or an insensitivity to it. Medical imaging is also helpful in diagnosing some endocrine disorders. CT and MRI scans, sonography, and scintillation counting are useful for detecting abnormalities in the size of endocrine glands and the presence of tumors or cysts.

This chapter focuses on disorders related to the pituitary gland, thyroid gland, and adrenal cortex. Endocrine diseases and disorders associated with other body systems are presented in chapters 25 (digestive system), 26 (nutrition and metabolism), 27 (male reproductive system), and 28 (female reproductive system).

Disorders of the Pituitary Gland

The pituitary gland secretes more hormones than any other endocrine gland and has a wider range of effects on the body. Therefore, pituitary disorders can produce an especially wide range of diseases. Here we consider only four of them: hypopituitarism, gigantism, acromegaly, and syndrome of inappropriate antidiuretic hormone secretion (SIADH).

Hypopituitarism

Hypopituitarism is a deficiency (hyposecretion) of some or all of the hormones normally produced by the anterior pituitary. *Primary hypopituitarism* is due to a defect of the pituitary gland itself, and *secondary hypopituitarism* to a failure of the hypothalamus to secrete the releasing factors that stimulate the pituitary. These two forms are usually indistinguishable, however, because it is not possible to measure the output of releasing factors from the hypothalamus to tell whether they are being produced at normal levels.

Hypopituitarism can result from pituitary infarction (pituitary necrosis caused by lack of blood supply, sometimes occurring in stroke, sickle-cell disease, and diabetes mellitus), head trauma, infections (such as meningitis and syphilis), surgical accidents, and other causes.

Regardless of the cause, the signs and symptoms of hypopituitarism depend on the hormones affected. *Panhypopituitarism* is a complete lack of pituitary tropic hormones. Because the pituitary gland then fails to stimulate the other endocrine glands "downstream" from it, those glands fail to secrete their own hormones as well. Panhypopituitarism therefore triggers a broad spectrum of dysfunctions resulting from hyposecretion of the thyroid, adrenal glands, and gonads and a lack of growth hormone. The signs and symptoms of panhypopituitarism include vomiting, anorexia, fatigue, loss of body hair, amenorrhea (cessation of menstruation) in women, testicular atrophy in men, decreased libido (sex drive), decreased metabolic rate, and reduced cold tolerance. The lack of ACTH and resulting hyposecretion of cortisol are especially life-threatening.

Women who have just given birth are particularly vulnerable to panhypopituitarism. During pregnancy, the pituitary gland increases markedly in size, metabolic rate, blood flow, and oxygen demand. If a woman experiences circulatory failure for any reason, the arteries of the pituitary gland, like arteries elsewhere, may constrict to compensate for falling blood pressure. If the pituitary suffers a circulatory deficiency for more than several hours, it begins to exhibit tissue necrosis and a chain of events that can lead to irreversible pituitary damage.

The diagnostic signs of panhypopituitarism include an enlarged sella turcica visible on X ray and low blood hormone levels detected by a technique called **radioimmunoassay,** in which radioactively labeled antibodies bind to any hormone that is present. One treatment option is **hormone replacement therapy (HRT),** in which the missing hormones are taken orally or by injection. Replacement of cortisol and thyroid hormone is essential, whereas the sex steroids may be replaced or not according to the patient's needs and wishes. Correction of cerebral edema sometimes restores ACTH and cortisol secretion.

Gigantism and Acromegaly

Hypersecretion of growth hormone (GH) causes **gigantism** if it begins in childhood and **acromegaly** if it begins or continues in adulthood. GH hyposecretion most often begins between the ages of 20 and 50, so acromegaly is the more common of the two disorders. Yet even this is rare, affecting only 1 person in 25,000. Most people with GH hypersecretion are first diagnosed in their 40s or 50s even though the disorder has, by then, been years in the making.

GH hypersecretion usually results from a benign, slow-growing tumor called a *pituitary adenoma.* In some cases, however, it is triggered by ectopic GHRH secretion, most often by tumors in the lung or pancreas. The excess GH stimulates increased connective tissue growth and disrupts carbohydrate metabolism. The connective tissue growth results in thicker and darker skin, coarsened body hair, and elongation or deformity of the hands, feet, ribs, mandible, and forehead, eventually resulting in coarsening of the facial features. The voice deepens because of increased connective tissue in the larynx. An increase in the number, size, and activity of sweat and sebaceous glands may cause profuse perspiration and disagreeable body odor. Peripheral nerves often become compressed, leading to neuropathies. Most internal organs increase in size, resulting in organ compression and pathological changes, some of which are not yet fully understood.

The disruption in carbohydrate metabolism results in **hyperglycemia,** an elevated blood glucose level. The pancreatic islets respond to this with insulin hypersecretion. This in turn stimulates target organs to down-regulate their insulin receptors, thus leading to insulin resistance and diabetes mellitus, seen in about 1 in 6 people with GH hypersecretion.

When GH hypersecretion begins in childhood or adolescence, before the epiphyseal plates of the long bones have closed, it stimulates bone elongation and gigantism. People with gigantism also show swelling of the soft tissues, enlargement of the nerves, delayed puberty, and often persistent sexual immaturity.

Besides the characteristics just described, there are other diagnostic signs of GH hypersecretion. Further tests to confirm the diagnosis often reveal poor glucose tolerance; X rays show an enlarged sella turcica and frontal sinuses, thickened skull, and deformities of the phalanges; a CT scan can show a pituitary adenoma; and radioimmunoassay shows an elevated blood GH concentration. If an adenoma exists, the syndrome can be treated by surgical **ablation** (removal) of the tumor, entering the cranial cavity through the sphenoid sinus, or ablation of the tumor with high-intensity radiation. Either form of ablation is a delicate procedure requiring care not to damage the hypothalamus and cranial nerves or to throw the patient into the opposite condition, hypopituitarism. Drug therapy lowers GH secretion in some patients. If left untreated, acromegaly progresses slowly but decreases life expectancy.

Syndrome of Inappropriate ADH Secretion

Disorders of the posterior pituitary gland are relatively rare, and when they occur, they usually involve antidiuretic hormone (ADH). ADH is normally secreted in response to dehydration, and its secretion is inhibited when the body is well hydrated and the blood is isotonic or hypotonic. However, **syndrome of inappropriate ADH secretion (SIADH)** is a state in which there is excess ADH secretion for the body's state of hydration. Usually, SIADH results from ADH secretion by ectopic sites such as tumors of the lung, duodenum, pancreas, and other sites. It also occurs in connection with some pulmonary infections and psychiatric diseases.

SIADH causes an increase in body water content and a loss of sodium, thus resulting in hyponatremia (sodium deficiency) and hypotonic hydration (excessively low osmolarity of the body fluids).. Since the action potentials in nerve and muscle physiology depend on sodium, hyponatremia can cause severe neurological and muscular dysfunction. The signs and symptoms of SIADH include anorexia (loss of appetite), impaired taste, dyspnea (difficulty breathing), fatigue, abdominal cramps, vomiting, confusion, muscle twitching, and convulsions. Despite being overly hydrated, a person with SIADH feels very thirsty.

SIADH is diagnosed through blood and urine tests that reveal hyponatremia, hypotonic blood serum, and hypertonic urine. Sodium intake and excretion are equal in SIADH, in contrast to other disorders (such as profuse sweating or aldosterone hyposecretion) in which hyponatremia may result from sodium excretion in excess of intake.

The most important priority in treating SIADH is to restore normal Na^+ concentration by means of I.V. fluid therapy and restriction of oral fluid intake. Most patients recover within 3 days. However, it is necessary to correct the underlying cause if any permanent cure is to be achieved. A drug called phenytoin helps to control excessive ADH secretion by the pituitary, but there is no drug available for stopping ectopic ADH secretion. Ectopic ADH secretion requires elimination of the underlying cause such as a tumor or a lung infection.

Disorders of the Thyroid Gland

Hyperthyroidism is the excessive secretion of thyroid hormone (TH), and **hypothyroidism** is inadequate TH secretion by the thyroid gland itself. Remember that there are three levels of control over TH secretion: (1) The hypothalamus secretes thyrotropin-releasing hormone (TRH); (2) TRH stimulates the anterior pituitary to secrete thyrotropin, or thyroid-stimulating hormone (TSH); and (3) TSH stimulates the thyroid gland to secrete TH. Consequently, there are at least these three places where something can go wrong with TH secretion. A disorder of TH secretion is called *primary* when it results from a disorder of the thyroid gland itself, *secondary* when it results from a disorder of TSH secretion by the pituitary, and *tertiary* when it results from a disorder of TRH secretion by the hypothalamus.

Hyperthyroidism

The most common cause of primary hyperthyroidism is **Graves disease (toxic goiter),** which is thought to be a hereditary autoimmune disorder. More than 95% of patients have immunoglobulins, termed *thyroid antibodies*, which directly stimulate the thyroid gland to secrete TH. In extreme cases, damage to the retina and optic nerve may occur, resulting in irreversible blindness. Hyperthyroidism is also induced by elevated TRH or TSH secretion, thyroid cancer, and some other conditions.

All forms of hyperthyroidism, however, produce similar signs and symptoms affecting virtually every organ system. These include an enlarged thyroid gland (goiter), anorexia, weight loss, diarrhea, nausea, vomiting, heat intolerance, excessive sweating, flushed and warm skin, eyelid tremor, tachycardia (abnormally fast heartbeat), loud heart sounds, hypertrophy of the left ventricle of the heart, restlessness, fatigue, insomnia, decreased attention span, and dyspnea. Hyperthyroidism is diagnosed from blood tests that show elevated TSH or TH levels, a test of thyroid function that shows an elevated uptake of radioactive iodine, and a thyroid scan.

Treatment is aimed at controlling the secretion of TH. This can be done in a number of ways depending on the source of the elevated TH. Common treatments for primary hyperthyroidism include the ablation of thyroid tissue by surgery or radiation and drug therapy to reduce TH secretion. If the thyroid gland is destroyed, the patient must be placed on hormone replacement therapy (HRT) to prevent hypothyroidism.

Hypothyroidism

Hypothyroidism is the most common disorder of thyroid function. Primary hypothyroidism is more common than the secondary or tertiary forms and most frequently results from thyroiditis, inflammation of the thyroid. *Acute hypothyroidism* is rare and is caused by bacterial infection. *Subacute hypothyroidism* is a nonbacterial inflammation of the gland, often preceded by a viral infection. These disorders are characterized by fever and an enlarged, tender thyroid. They last for 2 to 4 months, are treatable with corticosteroids, and produce no lasting damage. *Autoimmune thyroiditis (Hashimoto disease),* the most common form of primary hypothyroidism, results from the destruction of thyroid tissue by autoantibodies. It causes permanent damage to the thyroid gland and requires lifelong hormone replacement therapy with oral TH.

As with hyperthyroidism, hypothyroidism affects virtually every body system. Signs and symptoms include confusion, memory loss, lethargy, slow speech, cerebellar ataxia, decreased libido and sexual function, anemia, decreased heart rate, cool skin, cold intolerance, dyspnea, and reduced renal blood flow. Patients may also exhibit weight gain, constipation, fluid retention, elevated blood lipids, slow movements, dry and flaky skin, brittle hair, myxedema, and aching and stiffness in muscles and joints. Children with hypothyroidism show delayed development and mental retardation.

Hypothyroidism is diagnosed by the same techniques as hyperthyroidism. Blood tests reveal abnormally low TH levels. Primary, secondary, and tertiary hypothyroidism are differentiated by testing the patient's response to TRH and TSH. Primary hypothyroidism is characterized by a lack of response to both TSH and TRH, while in secondary hypothyroidism there is response to TSH but not to TRH. Tertiary hypothyroidism is defined by responsiveness to either TRH or TSH. Treatment consists of TH replacement to attain normal levels. The amount of TH given depends on the patient's age, the duration and severity of the hypothyroidism, and the presence of any other complicating conditions.

Disorders of the Adrenal Cortex

The adrenal cortex synthesizes more than 25 hormones, known collectively as corticosteroids, or corticoids. Many disorders can result from adrenal hypersecretion or hyposecretion, and their physical effects vary widely.

Cushing Syndrome

Cushing syndrome is the chronic hypersecretion of cortisol by the adrenal cortex, regardless of cause. When it results from ACTH secretion by the anterior pituitary, it is called **Cushing disease,** but Cushing syndrome can also result from adrenal tumors that secrete cortisol and from ectopic ACTH-secreting tumors elsewhere in the body.

People with Cushing syndrome commonly exhibit acne, hyperpigmentation and thinning of the skin, hirsutism (increased body and facial hair), obesity of the trunk, transient weight gain, a pendulous abdomen, flushed face, "buffalo hump," "moon face," increased susceptibility to bruising, muscle wasting, and weakness in the limbs. The chronically high cortisol level promotes hypertension, suppresses the immune system and wound healing, and makes a person highly susceptible to infections. About half of patients experience psychological effects, including irritability, depression, and schizophrenia. People with Cushing syndrome can lead normal lives with treatment, but without treatment, about 50% die within 5 years of onset. The most common causes of death are overwhelming infection, hypertension, arteriosclerosis, and suicide.

Cushing syndrome is diagnosed through physical examination and laboratory tests to measure the concentrations of ACTH and cortisol in the blood, the amount of cortisol in the urine, and the ability of dexamethosone (a synthetic glucocorticoid) to suppress ACTH secretion. Cushing syndrome is indicated when ACTH and cortisol are elevated and dexamethosone is unable to suppress ACTH. Treatment is designed to decrease cortisol production and is therefore specific to the underlying cause of the hypercortisolism. Treatment approaches include medication, radiation, and surgery.

Addison Disease

The hyposecretion of cortisol can result from *primary adrenal insufficiency*—inability of the adrenal cortex to secrete cortisol—or from inadequate stimulation of the adrenal cortex by ACTH. Primary adrenal insufficiency is called **Addison disease.** This relatively rare disease (affecting 4 people per 100,000) has an autoimmune basis and strikes women more than men. Clinical signs of the disease do not appear until at least 90% of the adrenal cortex has been destroyed by autoantibodies, so the onset of disease is usually between the ages of 30 and 60.

In Addison disease, the level of ACTH is elevated, and the signs and symptoms result from this fact as well as from the deficiency of cortisol. Thus, some of the signs of Addison disease are like those of Cushing syndrome, while others differ. The signs include hyperpigmentation, weakness and fatigue, anorexia, weight loss, vomiting, diarrhea, hypoglycemia, mental confusion, hypotension, dehydration, elevated red blood cell count, and vitiligo (white patches of depigmented skin).

Diagnosis is achieved through physical examination and laboratory tests. Decreased plasma and urinary cortisol indicate adrenal insufficiency. The condition is treated with daily hormone replacement therapy coupled with dietary modification. Hormone replacement strives to return circulating cortisol (and aldosterone if necessary) to near-normal concentrations. Patients must supplement the cortisol in times of stress to allow for normal physiological adaptation to stress. Their diets must ensure adequate sodium intake and compensate for sodium loss due to excessive sweating or diarrhea.

Case Study 17 The Woman with Weight Gain

Linda is a 52-year-old mother of two who was diagnosed with rheumatoid arthritis (RA) 8 years ago. Since then, she has been taking prednisone, a synthetic glucocorticoid, to minimize the inflammatory effects of the disease. Over the course of her treatment, Linda's physician has gradually had to increase her daily dose of prednisone.

Linda has developed a "pot-belly," a rounder face, and more facial hair. Concerned about these changes, she visits her physician, who completes an examination. While updating the patient history, the physician learns that Linda has also noticed that she bruises easily. The physician tells Linda that her symptoms are probably related to the use of prednisone rather than to another disease. Blood samples are drawn, and the physician asks Linda to return in a week.

At her next appointment, Linda is informed that, with the exception of RA, she is in good health, with no sign of any other disease. The physician tells her that the changes she is noticing will continue unless she stops or reduces her intake of prednisone. Also, additional effects of the prednisone could develop over time, such as acne, muscle wasting, thinning hair, and a slight hump on her back. After further discussion, Linda and her physician decide to gradually reduce the amount of prednisone she is taking and switch to other anti-inflammatory agents to control her RA symptoms.

Based on this case study and other information in this chapter, answer the following questions.

1. The signs caused by Linda's use of prednisone mimic those of an endocrine disease. Identify that disease.
2. What signs or symptoms of that disease would you predict that Linda will *not* exhibit? Why not?
3. What other side effects would you expect prednisone to have? Explain them.
4. Based on your understanding of the regulation of hormone receptors, explain why the physician would recommend a *gradual* withdrawal of prednisone and not an abrupt cessation of its use.
5. Describe how prednisone would affect the secretion of ACTH and cortisol.
6. Based on what you know of the network of controls over endocrine gland function, explain why the ACTH level is *elevated* in Addison disease.
7. It was explained in this chapter that people with growth hormone hypersecretion may develop diabetes mellitus as a side effect. Would you expect this diabetes to be type I (IDDM) or type II (NIDDM)? Explain your reasoning.
8. Explain why people with growth hormone hypersecretion may exhibit an enlarged sella turcica on X ray.

9. An asthma patient receives a prescription for theophylline, a phosphodiesterase inhibitor used as a relaxant for smooth muscle. How would you expect this drug to affect cellular responses to hormones that activate adenylate cyclase? Why? If caffeine also acts as a phosphodiesterase inhibitor, what do you think the doctor would recommend to this patient regarding caffeine consumption?

Selected Clinical Terms

ablation The removal of tissue by surgery, radiation, or other means.

ectopic In an abnormal location, such as a tumor in a nonendocrine organ that secretes the same hormone as an endocrine gland.

hormone replacement therapy (HRT) The administration of a hormone orally or by injection to compensate for insufficient endogenous production of the hormone—for example, taking thyroid hormone pills to compensate for thyroid hyposecretion.

hypersecretion The secretion of excessive amounts of a hormone.

hyposecretion The secretion of an insufficient amount of hormone to maintain normal homeostasis.

radioimmunoassay A way of testing for the presence of a hormone or other chemical by introducing a radioactively labeled antibody that binds to that substance; the concentration of radioactivity in a tissue then indicates the presence of the hormone or other ligand for that antibody.

18 The Circulatory System: Blood

Objectives

In this chapter we will study

- how blood and bone marrow are collected and clinically analyzed;
- some risks and benefits of blood transfusion;
- some methods that athletes use to increase their hematocrits, endurance, and competitiveness, and the risks and legitimacy of these methods; and
- several diseases affecting the blood—iron overload, anemia, leukemia, and thrombocytopenia.

Evaluation of the Blood and Clotting Systems

The blood plays several important roles in the body, and throughout this manual we have seen that it provides clues to disorders involving multiple body systems. Therefore, it is not surprising that blood tests are the most common laboratory tests performed in conjunction with a physical examination. This chapter focuses on how blood samples are obtained and analyzed and then discusses some diseases that affect the blood.

Collecting Blood Samples

Blood is most often collected by **venipuncture**—that is, drawn from a vein through a needle into a vacuum tube. A health-care professional who specializes in drawing blood for clinical purposes is a **phlebotomist.** Blood collection tubes are usually pretreated with an anticoagulant such as ethylenediaminetetraacetic acid (EDTA), a chemical that binds calcium ions and blocks the clotting cascade. Other anticoagulants include heparin and citric acid. When serum is required, anticoagulants are omitted from the tube, the blood is allowed to clot, and the clot solids are separated from the serum. When small amounts of blood are needed or when venipuncture is not feasible, blood may be obtained by using a sterile lancet to puncture the earlobe or finger in adults or the heel in infants. The blood may then be collected in a capillary tube.

Analysis of Blood Properties

After collection, blood may be evaluated for various purposes. For example, biochemical examination tells the clinician whether the body is adequately supporting hemopoiesis and reveals the amount of iron, total iron-binding capacity, and concentrations of vitamin B_{12}, folic acid, and hormones that stimulate blood cell development. A number of tests are available to examine the condition of erythrocytes, leukocytes, platelets, and clotting factors. Such observations, taken in conjunction with previous tests, are often crucial in diagnosing blood disorders.

A **complete blood count (CBC)** provides the number of erythrocytes (RBCs), leukocytes (WBCs), and platelets per microliter of blood; a differential WBC count (the percentage of the total WBC count composed of each WBC type); hematocrit (Hct); and erythrocyte morphology. Erythrocyte morphology is described in terms of RBC size, shape, and pigmentation (hemoglobin content).

Blood cells used to be diluted in a pipet and visually counted under a microscope, but this process was time-consuming and gave highly variable results. Most laboratories now use electronic instruments such as a Coulter counter, which forces blood cells through a narrow orifice and counts them as voltage pulses that vary with the cell type and size. Such **electronic cell counters** give much faster and more accurate readings and are able to distinguish RBCs, platelets, and different types of WBCs from each other, and to measure hemoglobin, hematocrit, and the **erythrocyte indices** shown in the following list. The first of these is the most valued and trusted datum and is determined by electronic cell counters. The others are calculated from the given data, but are not as reliable. Note that the term *corpuscular* in these indices refers to the fact that RBCs are also sometimes called *red blood corpuscles.*

- *Mean corpuscular volume (MCV)* is the average volume of the RBCs, measured in cubic micrometers (μm^3) or the equivalent unit, femtoliters (fL). The normal MCV is $90 \pm 9\ \mu m^3$.

Microcytosis is an MCV < 80 μm^3, and *macrocytosis* is an MCV > 100 μm^3.

- *Mean corpuscular hemoglobin (MCH)* is the average weight of hemoglobin per RBC, measured in picograms (pg). The normal MCH is 32 ± 2 pg.

- *Mean corpuscular hemoglobin concentration (MCHC)* is the average percentage concentration of hemoglobin in the RBCs. The normal MCHC is 33 ± 3%. *Normochromia* is an MCH and MCHC within the normal ranges given here, while *hypochromia* (abnormally pale RBCs) is defined as either an MCH < 27 pg or an MCHC < 30%.

Some additional abnormalities in RBC appearance are *anisocytosis,* abnormal variation in cell size, and *poikilocytosis,* abnormally irregular shapes. *Reticulocyte count* is also important. Reticulocytes are immature RBCs that have been recently released by the bone marrow. They still possess a fine network of endoplasmic reticulum. Normally, about 1% of the circulating RBCs are reticulocytes. An elevated reticulocyte count suggests abnormally rapid bone marrow activity (accelerated erythropoiesis), which sometimes indicates that the body is compensating for a condition such as anemia or *hypoxemia* (low blood O_2 level).

If these tests do not supply enough information to complete a diagnosis or if the information from the CBC suggests other disorders, additional tests may be run, including coagulation tests and analysis of the bone marrow.

Coagulation Tests

If a clotting disorder is suspected, various laboratory tests may be performed to measure the effectiveness of coagulation. Two of these are *bleeding time* and *prothrombin time.*

- **Bleeding time** is the length of time required for bleeding from a defined puncture to cease. For example, when the earlobe is punctured with a lancet, it should stop bleeding in 1 to 3 minutes.

- **Prothrombin time (PT)** is measured by isolating a sample of the patient's plasma, treating it with citrate to bind all the calcium, adding thromboplastin, then adding excess calcium to "recalcify" the plasma and measuring the time from then to clot formation. The blood should clot within 12 seconds, but takes longer if there is a deficiency of prothrombin, fibrinogen, or factors V, VII, or X.

The etiology of clotting disorders revealed by these tests can be narrowed down by means of additional coagulation tests.

Bone Marrow Analysis

Bone marrow is examined to determine whether hemopoiesis is taking place normally. Under local anesthesia, a small (25 μL) sample of marrow is removed by aspiration (suction) from the sternum, iliac crest, or spinous processes of the vertebrae. The marrow is smeared on a microscope slide, stained, and examined for level of hemopoietic activity, ratio of myeloid to erythroid elements (M/E ratio), number of megakaryocytes, maturational abnormalities, presence of cancerous cells, and relative frequency of different cell types. If an infectious disease is suspected, some of the sample may be cultured in order to identify the infectious microorganism.

Blood Transfusions

In the United States alone, as many as 15 million transfusions are performed each year. As with any medical procedure, blood transfusions involve risks that the patient must be made aware of. Complications that can develop during or following a transfusion include disease transmission, hemolytic reactions, and allergic reactions. Diseases that can be transmitted through transfusion include malaria, bacterial infections such as syphilis, and viral infections such as hepatitis and AIDS. Hemolysis most often results from an incompatibility of blood types, a risk that is minimized by proper cross-matching of donor and recipient blood. Allergic reactions can occur if the donor plasma contains something to which the recipient is allergic.

Nevertheless, the benefits of transfusion generally outweigh the risks, especially since strict screening procedures for donors and donated blood have been established. Would-be donors are disqualified if they have a history of hepatitis, heart disease, bleeding disorders, HIV infection, hypertension, or anemia. Hepatitis is grounds for permanent disqualification, whereas recent tattooing or body piercing may temporarily disqualify a donor. Donation is usually limited to once every 2 months to prevent donors from developing anemia.

In a routine blood donation, about 500 mL of blood are collected under sterile conditions into a plastic bag, usually containing citrate and dextrose. The blood may be stored as whole blood or separated into various components such as packed RBCs, plasma, and platelets. Whole blood can be stored for up to 35 days, whereas blood fractions can be stored longer.

The rising prevalence of infectious diseases has increased the demand for **autologous transfusions** in which a patient who anticipates surgery donates some of his or her own blood, which is stored and reinfused into that person when needed. Advances in technology also now enable blood shed during surgery to be collected and transfused back into the patient later.

Artificial Manipulation of the Hematocrit

Endurance athletes such as swimmers and runners depend on the oxygen transport capacity of the blood to supply their muscles, and the oxygen transport capacity is determined by the quantity of RBCs and hemoglobin in the blood. Some athletes employ a questionable technique called **blood doping** to raise their hematocrits and thus their endurance. Blood is removed weeks before a competition and separated into plasma and RBCs. The plasma is immediately reinfused, while the RBCs are stored. During the ensuing weeks, the body regenerates most of the lost RBCs and returns the hematocrit to normal. Then, a few days prior to competition, the stored RBCs are reinfused, giving the athlete an abnormally high RBC count. Although this technique increases endurance, the resulting polycythemia stresses the heart and blood vessels. Some athletes now choose an alternative to blood doping—using recombinant human erythropoietin (EPO) to stimulate erythropoiesis. The International Olympic Committee and numerous other amateur and professional sporting organizations have banned both blood doping and EPO use.

Another factor that elevates the hematocrit is living at high altitudes (10,000–12,000 feet). This occurs because hypoxia due to the decreased atmospheric oxygen stimulates increased EPO secretion by the kidneys. Knowing this, some endurance athletes train at high altitudes for competitions to be held at lower altitudes, temporarily giving themselves increased low-altitude endurance.

Iron Overload

Iron overload, or **hemosiderosis,** is the deposition of excess iron in the tissues in the form of *hemosiderin,* a hemoglobin breakdown product. It occurs in people who undergo repeated transfusions, have hemolytic anemia, take in excess iron, or drink excessive amounts of alcohol. Normally, the body contains about 5 g of iron, but in hemosiderosis, this can rise as high as 80 g. The lungs and kidneys are especially subject to iron accumulation. **Hemochromatosis** is a hereditary form of iron overload in which the body absorbs and stores too much iron from the diet. Iron overload leads to cirrhosis of the liver, bronzing of the skin, and sometimes diabetes mellitus, pituitary failure, and heart failure. The treatment of hemosiderosis is aimed at reducing the amount of iron in the body. This can be done by removing about 500 mL of blood (which contains 250 mg of iron) per week until serum iron levels return to normal, then removing 500 mL once every 3 to 4 months to maintain the appropriate level. In some countries, iron overload is treated by oral medications called **chelating agents** that bind iron (ferrous ions, Fe^{2+}) and make it unreactive until it is excreted from the body. (Similarly, EDTA, mentioned earlier in this chapter, is a chelating agent for Ca^{2+}.)

Anemia

Anemia, a deficiency of either red blood cells or hemoglobin, is most often the result of an iron deficiency. **Iron deficiency anemia** is the most common type of anemia worldwide. In developing countries, the most common cause is infection with hookworms—roundworms that attach to the intestinal wall and suck blood from the mucosa. In developed countries, the most common causes are chronic blood loss (for example, from a bleeding ulcer) and pregnancy. In the United States, iron deficiency anemia occurs in about 20% of all females, 50% of pregnant females, 20% of preschool children (whose rapid growth may outstrip their iron supply), and 3% of males.

The symptoms of iron deficiency anemia develop slowly, and people usually do not seek treatment until their hemoglobin level drops below 7 or 8 g/dL (the norm is 12 to 18 g/dL). The symptoms include fatigue, weakness, and shortness of breath, and signs include paleness of the skin and conjunctivae. The RBCs exhibit microcytosis, hypochromia, anisocytosis, and poikilocytosis. The reticulocyte count is abnormally low. As the disease progresses, the nails become thin, brittle, concave or spoon-shaped, and marked with coarse longitudinal ridges. Soreness of the tongue and soreness and dryness of the corners of the mouth are also common. Iron is needed not only for hemoglobin but also for synthesis of myoglobin, cytochromes, and some enzymes. Iron deficiency thus affects more than the blood, and may lead to headache, irritability, pica (compulsive chewing on nonnutritive substances such as clay or ice), numbness, mental confusion, and memory loss. In the elderly, the signs of iron deficiency are all too often dismissed as "normal" signs of aging. Iron replacement for 1 or 2 weeks often reverses the signs and symptoms of iron deficiency anemia, but must be accompanied by measures to either rule out or correct blood loss.

Most cases of hemolytic anemia are either drug-induced or caused by autoantibodies that lyse a person's own RBCs. The latter disorder is called **autoimmune hemolytic anemia (AHA).** The signs and symptoms of AHA depend on the extent of the hemolysis versus the effectiveness of compensatory erythropoiesis. The spleen enlarges **(splenomegaly)** as it disposes of dead and defective RBCs. Jaundice occurs if the liver cannot dispose of bilirubin as fast as hemoglobin breakdown produces it. In severe cases, the proliferation of bone marrow "trying" to compensate for the RBC loss may cause bone deformities, weakness, and pathological fractures. The reticulocyte count is elevated, but the RBCs are normal in size and hemoglobin content. AHA is treated with folic acid to meet the demands on the high rate of RBC turnover, steroids to control the immune attack on the RBCs, and splenectomy if the spleen is overly enlarged and acting as the major site of RBC destruction.

Leukemia

There are many kinds of leukemia, but the predominant characteristic of all of them is an elevated concentration of circulating WBCs. Leukemia is a form of cancer and is thus caused by genetic defects in the hemopoietic tissues. It sometimes runs in families, and it can be induced by mutagens such as ionizing radiation, viruses, and drugs. Leukemia is classified as *acute* or *chronic*, depending on the time course of the disease, and as *lymphoid* or *myeloid,* depending on the hemopoietic tissues and WBC types involved.

Adults develop leukemia more often than children. More than two-thirds of childhood leukemia is the *acute lymphoblastic* form (named for the fact that lymphocyte development is arrested at the lymphoblast stage. In adults, acute nonlymphoblastic and chronic lymphocytic leukemias are more prevalent. Worldwide, the countries with the highest leukemia incidence are Canada, New Zealand, Sweden, and the United States.

On average, only 38% of patients with leukemia survive for 5 years. This low number is due to the especially low survival rate of those with acute leukemias. However, improved chemotherapy and other treatment techniques have significantly increased survival rates. In children with acute lymphoblastic leukemia, for example, survival improved from 4% in the 1960s to 80% by 1994.

Acute leukemias have a rapid onset and are characterized by the presence of primarily undifferentiated or immature WBCs. Chronic leukemias progress more slowly than acute leukemias and are characterized by WBCs that appear normal but function abnormally. The bone marrow of leukemic patients produces insufficient RBCs and platelets, and leukemic patients are therefore subject to fatigue (from anemia), bleeding disorders (from platelet deficiency), and infection (from lack of competent WBCs). Neurological disturbances commonly occur as leukemic cells invade the central nervous system, where the blood-brain barrier protects them from chemotherapeutic drugs. The neurological signs and symptoms include headache, blurred vision, hearing problems, and facial palsy.

The most common treatment for leukemia is chemotherapy in conjunction with supportive therapies such as blood transfusions, antibiotics, and antifungal and antiviral medications. Bone marrow transplants and immunotherapy to induce differentiation of immature granulocytes are being used with some success.

Thrombocytopenia

Thrombocytopenia is deficiency in the number of platelets (less than 100,000/μL). The decreased platelet count may arise from decreased production, increased destruction or utilization, dilution, or uptake of platelets by the spleen. Patients with thrombocytopenia may exhibit multiple **petechiae** (see chapter 7 of this manual), small **ecchymoses** (bruises), mucosal bleeding, excessive bleeding after trauma or surgery, weight loss, fever, and headache. The severity of these symptoms depends on how much the platelet number is reduced. Diagnosis is made through patient history, physical examination, the CBC, and various clotting tests.

Thrombocytopenia can be caused by inadequate nutrition, transfusion with packed RBCs, chemotherapy, viral infections such as HIV and cytomegalovirus (CMV), or certain drugs such as heparin. After the cause has been identified, treatment may include transfusion of platelets, glucocorticoids to minimize platelet destruction, and administration of immunosuppressive drugs.

Case Study 18 The Bruised Boy

Eight-year-old Gene is taken to his pediatrician by his father. Over the past 3 months, Gene has been overly tired, and he appears pale most days. Gene's parents have become concerned because he has lost interest in outdoor activities. His father describes him as previously active and always eager to play outside with his friends. He also mentions that Gene has been getting more bruises and nosebleeds than usual, even though he is not as active as he once was. When questioned by the pediatrician, Gene admits to having headaches and describes soreness in his joints.

Physical examination reveals numerous petechiae and ecchymoses. Gene's oral temperature is 100°F. Palpation of the liver and spleen reveal no abnormalities. Gene's breathing and heart sounds are normal, as are his blood pressure and heart and respiratory rates. A blood sample is drawn for a CBC, and the results are as follows:

Hematocrit (Hct) = 36%

Hemoglobin (Hb) = 8 g/dL

RBC count = $2.5 \times 10^6/\mu L$

WBC count = 20,000/μL

Neutrophil count = 7,500/μL

Basophil count = 200/μL

Eosinophil count = 400/μL

Monocyte count = 1,800/μL

Lymphocyte count = 4,700/μL

Platelet count = 15,000/μL

Lymphoblast count = 46% of marrow cells

These test results lead the physician to a diagnosis of acute lymphoblastic leukemia (ALL). Treatment is initiated, including chemotherapy, blood transfusion, and drugs to minimize infection. The pediatrician and the oncologist working on Gene's case tell his parents that although the prognosis is not good, Gene's chances of survival are improved because the diagnosis has been made early in the course of the disease.

Based on this case study and other information in this chapter, answer the following questions.

1. What specific findings in Gene's CBC support a diagnosis of acute lymphoblastic leukemia (ALL)?
2. If a bone marrow biopsy were performed, what findings would confirm the diagnosis of ALL?
3. Why is Gene's condition considered acute rather than chronic lymphoblastic leukemia?
4. How would Gene's signs and symptoms differ if he had acute myeloid leukemia?
5. Gene's father is concerned about the possibility of blood transfusions transmitting infectious disease such as hepatitis or AIDS. If you were the physician, how would you try to allay the father's fears?
6. How has Gene's leukemia affected his erythropoiesis? Support your answer with data from his blood work.
7. Sarah complains of fatigue and muscle weakness. Her sister Julie mentions that she suffered similar symptoms before being diagnosed with anemia. Julie tells Sarah that taking iron supplements corrected her anemia, so Sarah starts taking iron supplements daily. However, 6 months later, Sarah notices that her skin is becoming darker and her joints are sore. Which of the following may account for these new symptoms?
   a. hemosiderosis
   b. hemolytic anemia
   c. thrombocytopenia
   d. aplastic anemia
   e. blood doping
8. In adults, why is bone marrow for biopsy taken from the sternum, iliac crest, or spinous processes of the vertebrae rather than other sites?

9. When chelating agents are used as anticoagulants, why is it vital that these agents bind divalent cations (ions with a 2+ charge)?

10. Why do you think iron deficiency anemia is so much more common in females than in males and in pregnant women than in other women?

Selected Clinical Terms

bleeding time The time required for bleeding to stop from a defined skin puncture, varying with the puncture site and method; serves as a measure of the effectiveness of clotting mechanisms.

chelating agent Any chemical that binds ions such as calcium (Ca^{2+}) or iron (ferrous; Fe^{2+}) ions and prevents them from acting physiologically; used for such purposes as preventing unwanted blood coagulation and treating iron overload and heavy metal poisoning.

complete blood count (CBC) A blood analysis that includes an RBC count, WBC count, differential WBC count, platelet count, hematocrit, and RBC indices (see also *erythrocyte index*).

ecchymosis A bruise; skin discoloration caused by a large area of hemorrhagic blood, having a blue-black color at first, later changing to greenish brown or yellow.

erythrocyte index Any measure of RBC shape, size, and pigmentation (hemoglobin content), including mean corpuscular volume, mean corpuscular hemoglobin, and mean corpuscular hemoglobin concentration.

prothrombin time A measure of clotting effectiveness in which a blood sample is treated with citrate to chelate its calcium and prevent clotting, thromboplastin is added, and then calcium is added to override the chelating agent and the time is measured from calcium addition to the first appearance of clotting.

venipuncture The puncturing of a vein for any purpose, usually to draw a blood sample; also called phlebotomy and performed by a phlebotomist.

19 The Circulatory System: The Heart

Objectives

In this chapter we will study

- tests commonly used in diagnosing cardiovascular disorders;
- the general symptoms and treatment of cardiac arrhythmias;
- cardiac inflammation (pericarditis, endocarditis, and rheumatic heart disease);
- cardiomyopathy; and
- myocardial infarction.

Diagnosing Cardiovascular Disorders

The diagnosis of cardiovascular diseases begins with the patient history and physical examination. A patient who complains of tightness in the chest, a burning pain worsened by coughing, difficulty breathing, weakness, lightheadedness, and fatigue may be at risk for one of many cardiovascular diseases, especially if he or she also has a family history of heart disease. Various aspects of cardiovascular function are routinely assessed in a physical examination. The pulse rate, strength, and rhythm are examined by palpation; the heart sounds are studied by auscultation with a stethoscope; and the blood pressure is measured with a sphygmomanometer.

If a cardiovascular disorder is suspected, further testing is warranted. Such tests include noninvasive and invasive techniques as well as blood analysis.

Noninvasive Tests

Reduced cardiac output affects the oxygen and glucose supply to all organs, and the central nervous system is among the most sensitive to such deficiencies. Therefore, one of the most obvious signs of cardiac dysfunction is the impairment of psychological and motor functions such as attention, consciousness, coherent thought and speech, pupillary reflexes, and visual gaze and tracking movements. Poor cardiac output also causes the blood hemoglobin to become dark red or violet in color. This effect is most easily seen in areas of the body that have a dense capillary network and thin epithelium. Thus, a cardiovascular assessment includes noting the color of the gums and other mucous membranes, conjunctivae, and nail beds. *Cyanosis,* or blueness of the membranes, suggests reduced cardiac function, although it also has other causes. Cyanosis therefore brings to mind several etiological hypotheses and requires further tests to narrow down its cause.

Palpation of the pulse and auscultation of the heart sounds are vital to any cardiovascular examination. More sophisticated techniques involve various forms of *cardiography,* the measurement and recording of cardiac functions. The best-known of these techniques is *electrocardiography,* the recording of an electrocardiogram (ECG). A *phonocardiogram,* or record of the heart sounds, is made by placing a microphone on the *precordium* (the chest wall anterior to the heart) and connecting it to an amplifying and recording instrument. An **echocardiogram** is similar in principle to a fetal sonogram. Oil is spread on the chest, and a device is placed against it that generates ultrasonic vibration and detects the echoes that come back from the heart and associated structures. The record obtained through the cardiac cycle gives information on cardiac anatomy and such functional characteristics as stroke volume and cardiac output.

Other specific diagnostic tests include the following:

- A **pulse tracing** is a record of the pulsation produced by blood flowing through a vessel. It is produced by placing a sensor over a blood vessel such as the common carotid artery and recording fluctuations in blood pressure over the course of the cardiac cycle. This method is used in conjunction with the ECG and phonocardiography to determine the timing of the various events in the cardiac cycle.
- A **Doppler study** is a technique for listening to the sounds of blood flowing in the vessels for evidence of obstructions to flow or valvular defects in the heart. The sounds are amplified by a microphone hand-held over the blood vessel.

- A **stress test** is a study of the ECG and blood pressure during exercise. A patient typically walks a treadmill until the maximum heart rate for his or her age and sex is reached or until he or she begins to show signs of cardiac distress such as chest or leg pain, extreme fatigue, or extreme dyspnea. The ECG and blood pressure are then examined in comparison to pre-exercise records for indicators of cardiovascular disease.
- A **chest X ray** is a routine part of a cardiac examination. It shows the size, contour, and position of the heart relative to surrounding structures. A sharper silhouette of the heart can be obtained if the patient first swallows a contrast medium such as barium, which makes the esophagus appear as a bright white background against which the image of the heart stands out.

Invasive Tests

Except for the barium swallow, the techniques just mentioned are considered noninvasive because nothing enters the body. Invasive methods entail more risk but can provide an overriding benefit in some cases by giving the diagnostician more detailed and specific information.

In **cardiac catheterization,** a catheter (a thin, flexible tube) is threaded into a blood vessel until it enters a heart chamber. The catheter may then be used to determine pressure within a heart chamber, to withdraw blood for measurement of blood oxygen level, or to introduce a contrast medium that enhances images of heart chamber function, valvular function, or the coronary arteries. Visualization of the coronary arteries is called **coronary angiography (arteriography).** A catheter is threaded from the femoral artery into the left ventricle, and a contrast dye is injected to allow filming of ventricular function for a few cardiac cycles. Then the catheter is pushed to the openings of the coronary arteries, and dye is injected into the arteries so that they can be visualized. This method is used primarily to evaluate atherosclerosis.

Other invasive methods include PET scans, injection of radioisotopes to localize myocardial infarcts or ischemic areas, and electrocardiography using electrodes introduced into the heart by way of a catheter to record from the AV bundle.

Blood Analysis

Blood samples also provide information about cardiac function. When cardiac muscle tissue is destroyed (as in myocardial infarction), enzymes and other cytoplasmic components leak into the blood. The enzymes of greatest interest to cardiac function are aspartate transaminase, creatine kinase (CK), and lactate dehydrogenase (LDH). All of these enzymes appear in the serum within hours of a myocardial infarction, and each enzyme is elevated at different times. Aspartate transaminase peaks in 12 to 24 hours and returns to normal in 2 to 7 days. CK peaks within 24 hours and returns to normal in 3 to 5 days. LDH rises in 12 to 24 hours, peaks within 72 hours, and returns to normal in 8 to 12 days.

Additionally, circulating sodium and potassium concentrations serve as markers in a manner similar to that observed in skeletal muscle. The ability of the heart to effectively pump blood is measured in part by the various blood gas measurements. Normal measurements for serum enzymes, ion concentrations, and blood gases are found in the Appendix of Normal Values at the end of this manual.

Cardiac Arrhythmia

Cardiac arrhythmia (dysrhythmia) is any disturbance of the normal heart rhythm. The signs and symptoms of arrhythmias vary from patient to patient, but in general they include palpitations (awareness of the heartbeat), dizziness and syncope (fainting), and diagnostic alterations in the ECG. However, because the patient may not experience a bout of arrhythmia while in the clinic, the physician may have the patient wear a monitor to record heart activity over a 24-hour period or longer.

Arrhythmia is treated with several techniques, including anti-arrhythmic drugs or a pacemaker. An important aspect of treatment is to reassure the patient and reduce anxiety, especially in cases that produce palpitations but pose no health risk. Precipitating factors such as exercise, alcohol, or caffeine may be identified and the patient encouraged to modify his or her behavior.

Anti-arrhythmic drugs are the most common treatment. Four classes of anti-arrhythmic drugs are available, with the choice determined by the type of arrhythmia and the side effects of the drug: Na^+

channel blockers (lidocaine, quinidine, encainidine), β-blockers (propranolol, atenolol), K^+ channel blockers (sotalol, aminodarone), and Ca^{2+} channel blockers (verapamil, diltiazem). Pacemakers are small, battery-powered devices with electrodes that stimulate the atria or ventricles in response to events in the heart. Today's pacemakers are programmable and can be used to regulate both tachycardia and bradycardia.

Inflammatory Heart Diseases

A wide variety of microorganisms can infect the tissues of the heart and trigger cardiac inflammation, or **carditis.** Three examples are explored here—*pericarditis, endocarditis*, and *rheumatic heart disease.*-

Pericarditis

Conditions elsewhere in the body often lead to disorders of the pericardium. For example, infection, connective tissue diseases, and radiation therapy commonly trigger **pericarditis,** inflammation of the pericardium. Pericarditis produces sudden chest pain that is worsened by breathing, often making a person think he or she is having a heart attack. Other symptoms include irritability, restlessness, malaise, and difficulty swallowing. Signs found upon examination include tachycardia, a low fever, and a raspy, sandpaper-like sound called a **friction rub** heard at the apex and left sternal margin of the heart. The friction rub occurs when the inflamed, roughened pericardial membranes rub against each other. Pericarditis is treated with rest, analgesics, and nonsteroidal anti-inflammatory drugs.

Pericarditis usually resolves by itself in time, but some cases are complicated by **pericardial effusion,** the seepage of fluid into the pericardial cavity. If the fluid accumulates slowly, the pericardium can stretch to accommodate it, but if it accumulates rapidly, it can cause *cardiac tamponade,* a compression of the heart that prevents it from filling completely and thus reduces the stroke volume. As little as 50 to 100 ml of fluid may induce serious tamponade. Cardiac tamponade can be detected from a condition called *pulsus paradoxus* in which the arterial blood pressure is more than 10 mm higher when the patient exhales than when he or she inhales. Echocardiograms are the most sensitive way of confirming cardiac tamponade. If serious, pericardial effusion is treated by pericardiocentesis—puncturing the pericardium and withdrawing the fluid.

Endocarditis

Endocarditis is inflammation of the endocardium, especially the heart valves. It usually results from infection with bacteria, viruses, fungi, or parasites—but most often, the streptococcus and staphylococcus types of bacteria. It is often triggered by mitral valve prolapse, implantation of artificial heart valves, long-term use of cardiac catheters, I.V. drug abuse, and cardiac surgery. Males are affected twice as often as females.

Pathogenesis begins when a heart valve or other area of endocardium is "prepared" by endothelial damage to support colonization by microbes. Platelets adhere to the damaged region and produce a thrombus that can then serve as a focus of bacterial adhesion. Microbes can invade the blood from such sources as upper respiratory or skin infections, bladder catheterization, or even dental cleaning. They adhere to the thrombus and begin to proliferate, so that within 24 hours, there develops a lesion of alternating layers of bacterial colonies and clotted blood.

Typical signs of endocarditis include fever, weight loss, night sweats, a heart murmur, and abnormalities in the erythrocytes, urine, and ECG. The diagnosis is confirmed by culturing bacteria from the blood and by echocardiography. The disease is treated with antibiotics, but repetitive episodes of endocarditis may damage the valves so extensively that they require surgical replacement. Prevention of endocarditis is the reason some people are given antibiotics prior to receiving dental work.

Rheumatic Heart Disease

Rheumatic fever is an inflammatory disease caused by the immune response to a certain class of streptococcal bacteria. If untreated, it can lead to **rheumatic heart disease,** a scarring and deformity of the heart. Rheumatic fever arises most often in children from 5 to 15 years of age, developing solely as a complication of a streptococcal throat infection. If the throat infection is treated within 9 days, rheumatic fever usually does not develop. But if treatment is delayed, the infection progresses to rheumatic fever in about 3% of cases. This disease gets its name not only because it produces a fever but also because bacterial antigens bind to receptors in the synovial joints and trigger an autoimmune response, leading to widespread joint pain, among other symptoms.

About 10% of children with rheumatic fever go on to develop rheumatic heart disease. This syndrome begins with carditis in all three layers of the heart wall, but the endocarditis is the most serious. Beadlike clumps of "vegetation" develop on the heart valves and chordae tendineae. These structures become scarred and constricted, valve cusps may adhere to each other, and eventually a patient can die of cardiomegaly (enlargement of the heart), defects in electrical conduction in the heart, and left heart failure.

The first priority in treating rheumatic heart disease is to inhibit the inflammation. Aspirin (salicylate) is the treatment of choice, but cases unresponsive to aspirin are treated with corticosteroids. Penicillin G or other antibiotics are used to prevent recurrence of the streptococcus infection. Severely damaged heart valves may require surgical replacement.

Cardiomyopathies

Cardiomyopathies are structural or functional abnormalities of the myocardium. Most cases are idiopathic, but some are triggered by infectious diseases, toxins, cancer chemotherapy, alcoholism, connective tissue disease, or nutritional deficiencies. The two most common cardiomyopathies are *dilated* and *hypertrophic*.

Dilated cardiomyopathy is characterized by dilation of the ventricle and loss of contractility, so that the end-diastolic volume becomes greater (more blood remains behind in the heart with each beat) and the stroke volume is severely reduced. Patients commonly experience dyspnea, fatigue, palpitations, dysrhythmia, and dizziness. Dilated cardiomyopathy is treated with digitalis to stimulate the heart, diuretics to promote water excretion and lower the blood pressure, and bed rest, sometimes for extended periods. The prognosis depends on the extent of myocardial damage. Deaths from dilated cardiomyopathy usually occur within 5 years of diagnosis and most commonly result from left-ventricular failure.

Hypertrophic cardiomyopathy is marked by thickening of the interventricular septum. It seems to have a genetic basis. The diseased heart may appear normal in size, but thickening of the septum reduces the capacity of the ventricles. The ventricles stiffen and exhibit reduced filling and output. Angina, dizziness, palpitation, and dysrhythmia are among the signs and symptoms. Beta-blockers such as propanolol sometimes reduce ventricular stiffness and improve ventricular filling and ejection. Some cases are treated by surgically removing part of the hypertrophied myocardium. The chance of long-term survival is good with appropriate management.

Hope is now available for some patients with cardiomyopathies through heart transplants. But because donor hearts are in short supply, transplants are usually limited to patients under the age of 50 years, and even then most patients selected for transplant die before a donor heart becomes available. In the United States, it is estimated that only 10% to 12% of all patients with cardiomyopathies receive hearts annually. Successful transplantation results in a 50% to 70% 5-year survival rate.

Myocardial Ischemia and Infarction

Coronary atherosclerosis may cause myocardial ischemia—a prolonged deficiency of blood flow to the cardiac muscle—and lead to *myocardial infarction (MI),* or heart attack. Frequently, an MI occurs because platelets aggregate on an atherosclerotic plaque in a coronary artery and form a thrombus (blood clot), which can build up rapidly and block the artery or break loose and block a smaller artery downstream. If half or more of the artery lumen becomes blocked, blood flow may be inadequate to meet the metabolic needs of the myocardium, especially when cardiac workload increases.

The myocardium can tolerate about 20 minutes of ischemia before tissue death begins. Within 8 seconds, the myocardial oxygen reserves are depleted and the muscle shifts to anaerobic fermentation. Fermentation generates hydrogen ions and lactic acid, lowering the tissue pH and contributing in multiple ways to cellular injury. The cells leak K^+, Ca^{2+}, and Mg^{2+}; their contractility declines; and the heart's pumping ability is compromised. MI triggers a strong inflammatory response leading, if the patient survives, to tissue repair by fibrosis. Thus, some people's hearts exhibit scar tissue that indicates earlier infarctions of which they may have been unaware.

The first symptom of an acute MI is often severe chest pain, commonly described as a heavy, crushing sensation, "like an elephant sitting on my chest." The pain often radiates to the neck, jaw, back, shoulder, and left arm. Yet, in some cases, pain is absent. Consequently, the MI may not be immediately diagnosed or a patient may even deny that he or she has a life-threatening condition requiring emergency care. Patient denial is a major factor in the delay of treatment for MI and thus a major factor in mortality; 50% of deaths from acute MI occur in the first 3 to 4 hours of onset.

Other signs and symptoms of MI include restlessness, pallor, apprehension, sweating, and cyanosis. The pulse is unusually fine and difficult to feel ("thready"). An ECG often reveals arrhythmia with abnormal Q waves, changes in the S-T segment, and inverted T waves. Myocardial enzymes are elevated (see previous discussion), and within 12 hours the WBC count rises.

Diagnosis is based on the signs and symptoms just mentioned, along with the findings of imaging techniques. Patients are admitted to the hospital where the cardiac rhythm and serum enzymes can be monitored. Treatment involves the prompt administration of aspirin to minimize blood clotting and thrombolytic drugs such as tissue plasminogen activator (TPA) or streptokinase, which break up blood clots that already exist and restore myocardial perfusion in about 3 minutes. Pain relief is usually accomplished by administering sublingual nitroglycerin, a coronary vasodilator. Acute coronary care is followed by bed rest, dietary modification, and a gradual return to normal activity.

Long-term survival depends upon many factors—degree of left-ventricular ischemia and dysfunction, age, diet, and potential for ventricular dysrhythmias. About 8% to 10% of those who suffer an acute MI die within 1 year, and most of these within 3 to 4 months. The most common cause of death is ventricular fibrillation.

Case Study 19 The Hard-Working Executive

Paul is 42 years old and the president of a small company. He is just beginning his daily workout at the local gym when he notices a slight tightness in his chest. As he continues riding an exercise bike, the pain becomes more severe and radiates to his left arm, shoulder, and jaw. Paul decides he is "overdoing it" and heads for the showers, intending to go back to work for the rest of the day and see how he feels tomorrow. After returning to his office, he begins to sweat and mentions to his secretary that he has an upset stomach. He thinks he might be coming down with the flu, but needs to get a few things done before going home.

Approximately 10 minutes later, Paul's secretary finds him unconscious on the floor of his office and calls an ambulance. When the paramedics arrive, they find that Paul is not breathing and has no detectable pulse. His skin is pale, cool, and clammy. They initiate cardiopulmonary resuscitation, reestablish breathing and a regular heartbeat, administer the clot-dissolving agent TPA, and begin transporting Paul to the hospital. On the way, Paul regains consciousness, and a paramedic gives him an aspirin to chew. They determine the following vital signs:

Heart rate = 50 beats/min and irregular

Blood pressure = 74/48 mmHg

Respiratory rate = 16 breaths/min and shallow

At the hospital, Paul is promptly placed on a cardiac monitor, blood is drawn for enzyme analysis, and I.V. propanolol is given. Once Paul is stabilized, he is transported to the cardiac care unit (CCU). He is given nitroglycerin for pain relief and nasal oxygen to maintain an adequate blood O_2 level; his ECG and blood pressure are closely monitored by the nursing staff.

Among the blood test results are the following:

pH = 7.18

Lactate = 42 mEq/L

Creatine kinase (CK) = 82 IU/L

Lactate dehydrogenase (LDH) = 130 IU/L

These results, coupled with the ECG, preadmission symptoms, and patient history, confirm that Paul has suffered a myocardial infarction. Paul is kept in the hospital for 7 days for monitoring and treatment. His cardiologist and his primary care physician call on Paul in the hospital. His primary care physician is well aware of Paul's history: He is divorced and overweight, smokes up to three packs of cigarettes per day, is under treatment for atherosclerosis, and has a family history of hypertension. Paul is frightened by his hospitalization and resolves to lose weight and quit smoking. His physician reinforces these decisions, prescribes a mild tranquilizer to relieve Paul's anxiety, and discusses Paul's rehabilitation with him. After 3 days of bed rest, Paul is encouraged to get up, rest in a chair, read, and walk to the bathroom as needed. He is released at the end of the week, but scheduled for frequent visits to his cardiologist for the next 6 weeks and counseled on gradual resumption of normal physical activity and on the issues of smoking, weight loss, work habits, and diet.

Based on this case study and other information in this chapter, answer the following questions.

1. What risk factors predispose Paul to a myocardial infarction?
2. Explain the physiological basis of Paul's elevated serum lactate and lactate dehydrogenase concentrations.
3. Explain why Paul is given an aspirin to chew on the way to the hospital.
4. If Paul's MI results from occlusion of the circumflex coronary artery only, where would you expect to find the myocardial lesion in a PET scan of the heart?
5. Explain why Paul is given propanolol in the hospital. What type of drug is this? How would it improve his condition?
6. Both nitroglycerin and streptokinase can restore perfusion of the myocardium through an obstructed coronary artery, but they do so in different ways. Contrast the mechanisms of these two drugs.
7. Why would you expect intravenous drug users to have a high incidence of endocarditis?
8. Cardiac tamponade restricts the filling and stroke volume of the right side of the heart before the left. Why do you think this is so?
9. Heart attack patients in a CCU are ideally kept in a private room and allowed few visitors for the first 2 or 3 days. The room should have a clock, calendar, and window to the outdoors, and the patient may be provided with light reading if he or she wishes, but it is better not to furnish a radio, television, or newspapers. Explain the reason for all of this.
10. At the age of 20, Oscar is diagnosed with insufficiency of the bicuspid valve. Which of the following could indicate a predisposition for this diagnosis?
    a. hypertrophic cardiomyopathy
    b. pericarditis
    c. a childhood history of rheumatic fever
    d. cardiac arrhythmia
    e. pericardial effusion

Selected Clinical Terms

cardiac catheterization Insertion of a narrow flexible tube (catheter) into the heart for blood sampling, pressure measurement, or dye injections.

cardiomyopathy Any structural or functional abnormality of the myocardium.

coronary angiography A method in which a contrast medium is injected into the coronary arteries and an X ray made to assess arterial occlusion or other abnormalities.

echocardiogram An ultrasonic scan of the heart for the purpose of assessing cardiac anatomy and function.

endocarditis Inflammation of the endocardium, usually as a result of bacterial or other infection.

friction rub A raspy sound heard near the apex and left sternal margin of the heart when inflamed pericardial membranes rub across each other.

pericardial effusion Seepage of fluid into the pericardial cavity, presenting a risk of cardiac tamponade.

pericarditis Inflammation of the pericardium usually triggered by pathologies elsewhere in the body.

pulse tracing A method of measuring pulsations in blood pressure in a particular blood vessel over the cardiac cycle.

rheumatic heart disease Scarring and deformity of the heart, especially the endocardium, as the result of an autoimmune response to a streptococcus infection.

stress test An evaluation of cardiovascular fitness by recording the ECG and blood pressure during a defined strenuous exercise such as walking a treadmill.

20 The Circulatory System: Blood Vessels and Circulation

Objectives

In this chapter we will study

- aneurysms;
- syncope as a symptom of certain cardiovascular disorders;
- orthostatic hypotension;
- Raynaud phenomenon and Raynaud disease;
- patent ductus arteriosus in infants and children;
- varicose veins; and
- cerebrovascular accidents, or strokes.

Aneurysms

An **aneurysm** is a weak point in a blood vessel or in the heart where a thin-walled, bulging sac forms and may eventually rupture. Any blood vessel may develop an aneurysm, but the aorta is the most susceptible because it is subjected to the highest blood pressure. An aneurysm frequently results from a combination of atherosclerosis, which weakens the vessel wall, and hypertension, which adds stress to the wall. However, aneurysms can also result from syphilis, collagen diseases, or aging. Cigarette smoking, heredity, and trauma are other risk factors for aneurysm.

A *true aneurysm* involves the weakening and bulging of all three layers of an artery—the tunica interna, tunica media, and tunica externa. A *false aneurysm* is one in which blood pools in the wall of a vessel and is held there by a clot. A *dissecting aneurysm* is a condition in which the tunica interna is torn and blood from the arterial lumen flows into the wall, pooling between the tunica media and tunica externa, thus separating (dissecting) these layers. Dissecting aneurysms tend to develop abruptly, usually in the descending aorta, and they cause intense "tearing" or "ripping" pain, especially in the precordial or interscapular areas.

Aneurysms usually develop slowly and sometimes grow huge without producing any symptoms. They can be seen by X ray, sonography, CT, and other imaging methods, and they produce such signs and symptoms as a cough from pressure on the trachea, dysphagia from pressure on the esophagus, hoarseness from pressure on laryngeal nerves, **hemoptysis** (spitting up blood), a feeling of abnormal abdominal pulsation, and pain in the sternum, ribs, or spine. Cerebral aneurysms (common in the cerebral arterial circle) cause neurological signs due to increased intracranial pressure.

Aneurysms can cause collateral damage to the brain, lungs, nerves, and other adjacent organs by putting pressure on them, but the principal danger of an aneurysm is hemorrhage. The detection of an aneurysm calls for careful management of blood pressure (to prevent hemorrhage) and surgical repair if warranted by the severity of the aneurysm and permitted by its location. (Cerebral aneurysms may be irreparable.) Small aneurysms are managed with medication, but aneurysms beyond a certain size—and all dissecting aneurysms—demand emergency surgical attention. The diseased segment of artery may require replacement by a synthetic graft.

Syncope

Syncope is fainting, a brief loss of consciousness. It usually results from a drop in cerebral perfusion, due to either venous pooling or reduced cardiac output. It can be induced by heart rates lower than 35 beats per minute or greater than 150 beats per minute because cardiac output is reduced in both cases. Hypovolemia and other causes of reduced venous return can also lead to syncope.The loss of consciousness is often preceded by dizziness, light-headedness, nausea, sweating (diaphoresis), or blurred vision. A person who has fainted typically exhibits shallow breathing, a weak pulse, and low blood pressure.

Syncope itself usually requires no treatment, since the person normally drops to a horizontal position and gravity restores the cerebral blood supply. Recovery can be promoted by raising the subject's legs—for example, by propping the feet on a stack of books—so that blood from the legs drains back to the heart and restores cardiac output. However, the underlying cause of syncope should be identified and corrected if it is not obviously something minor such as an emotional surprise or shock. Syncope can be a sign of cardiovascular disease, hypovolemia, autonomic dysfunction, or other disorders requiring medical attention.

Orthostatic Hypotension

Orthostatic (postural) hypotension is a rapid drop in blood pressure that occurs when a person stands up. Essentially, it results from failure of the baroreflex to compensate for the sudden downward gravitational pull on the blood, failure of valves in the lower limbs to close and prevent the downflow of blood, or both. As a result, cerebral perfusion drops, and the person may feel dizzy or even faint. Fainting and falling, in turn, presents a risk of bone fractures or other injuries. This is especially serious for the elderly. About 20% of randomly selected elderly people have orthostatic hypotension—fewer in physically active, community-dwelling elderly and more in institutionalized elderly. A hip fracture in old age can mean a long period of immobility or even death from pneumonia and other complications of immobility.

There are two types of orthostatic hypotension—acute and chronic. *Acute orthostatic hypotension* is temporary, occurs infrequently, and results from a sluggish baroreflex. The underlying causes include some medications (vasodilators and diuretics), prolonged immobility, physical exhaustion, and decreased blood volume. *Chronic orthostatic hypotension* is of longer duration and occurs more frequently in a given person. It may result from adrenal insufficiency, diabetes mellitus, metabolic disorders, diseases that decrease blood volume, and disorders of the nervous system such as intracranial tumors, diabetic neuropathy, or Guillain-Barré syndrome.

The signs and symptoms of orthostatic hypotension vary with the degree of the decreased cerebral blood flow. Mild to moderate decreases may induce faintness, light-headedness, dizziness, confusion, and blurred vision. More severe decreases may cause syncope or even convulsions.

Orthostatic hypotension is diagnosed by monitoring the blood pressure as body position changes. If hypotension develops upon standing and is relieved by reclining, the patient is diagnosed with orthostatic hypotension. The treatment depends on the underlying cause. For example, if the condition is due to venous pooling in the legs, fitted elastic hose may be used to increase venous return and decrease venous pooling. Patients are also advised to change posture slowly and to sleep with the head elevated. Adrenergic drugs can be used, but require careful attention due to their side effects on other organ systems.

Raynaud Phenomenon and Raynaud Disease

Some people experience occasional spasmodic contractions of the arteries of the digits, especially the fingers. These spasms result in pallor, numbness, and coldness of the digits, typically beginning at the tips of the fingers or toes and progressing proximally. The digits often appear cyanotic at first, and then as normal blood flow returns, they turn red and may exhibit throbbing and paresthesia. In prolonged or severe cases, these repetitive ischemic episodes can lead to brittle nails or even to gangrene of the digits, necessitating-amputation.

When these spasms occur for a known reason, the disorder is called the **Raynaud phenomenon.** Causes include collagen diseases, pulmonary hypertension, and long-term occupational exposure to such conditions as cold or vibrating machinery. **Raynaud disease** is similar in its symptoms and effects but is idiopathic, occurs especially in young women, and is often triggered by emotional stress or brief exposure to cold.

Attacks can often be prevented by swinging the arms back and forth to drive blood to the fingertips and by avoiding the situations that trigger attacks. Since Raynaud disease is idiopathic, no treatment is available. Extreme cases of Raynaud phenomenon can be treated by administering vasodilators or calcium channel blockers, or by severing the sympathetic nerves.

Patent Ductus Arteriosus

The *ductus arteriosus (DA)* is a short vessel in the human fetus that connects the pulmonary trunk to the ascending aorta. Since the lungs perform no respiratory function until after birth, it is pointless for the right ventricle to pump all of its blood through them. The ductus arteriosus allows blood from the right ventricle to bypass the lungs and go directly into the systemic circulation. After birth, inflation of the lungs and changes in blood pressure gradients cause reversal of the blood flow in the ductus arteriosus. The ductus constricts, closes off this connection within 10 to 15 hours, and becomes a fibrous cord, the *ligamentum arteriosum,* within 2 to 3 weeks.

In some infants, however, the ductus arteriosus fails to constrict. It remains open, or patent, thus presenting the condition called **patent ductus arteriosus (PDA).** Blood in the aorta (which has recently passed through the lungs) now shunts into the pulmonary trunk and travels to the lungs again. This increases pulmonary blood flow and pressure as well as the workload on the heart. PDA may be asymptomatic, but is often characterized by poor weight gain and frequent respiratory infections, and sometimes by left heart failure.

PDA is detectable from chest X rays showing enlargement of the left heart and a continuous, characteristic murmur heard on auscultation or detected by echocardiography. Indomethacin, a prostaglandin inhibitor, sometimes stimulates the DA to close in premature infants, but if this fails, surgical ligature and bisection of the DA is usually performed (that is, the DA is tied off and then cut). This procedure is virtually always successful and seldom has any fatal complications. Surgical closure of the DA is usually done between the ages of 6 months and 3 years.

Varicose Veins

A **Varicose vein** is a vein that has become distended by pooled blood. The most common sites are the great and small saphenous veins of the lower limb, but any vein can be affected. As a case in point, hemorrhoids are varicose veins of the anal canal. Varicose veins develop as trauma damages the venous valves, long periods of standing stretch the veins with pooled blood, or obesity or pregnancy compresses veins and blocks venous return. For example, when blood pools in the veins of the lower limb, the surrounding tissues become edematous, the veins become dilated and convoluted, and the skin of the foot and ankle becomes hyperpigmented by RBCs that leave the circulation and accumulate in the tissues. Circulation can become so poor that the lower limbs develop pressure ulcers.

Varicose veins are incurable, but they can be treated to improve their appearance and relieve the symptoms. The condition is treated conservatively if possible by having the patient wear lightweight compression hosiery and avoid–standing for long periods of time. If these measures are insufficient, heavier elastic support hose may be required. Pain, phlebitis (vein inflammation), and disfigurement may justify surgical treatment. **Vein stripping** is the surgical removal of the saphenous veins, while **sclerotherapy** is an alternative to surgery in which the varicose vein is injected with a chemical that obliterates it by inducing fibrosis. Both vein stripping and sclerotherapy are avoided unless the saphenous veins are varicosed all the way from groin to ankle. It is desirable to preserve the saphenous veins if possible, because they are the best source of vein in the event that the patient ever needs coronary bypass surgery.

Cerebrovascular Accident (Stroke)

Stroke is known clinically as a **cerebrovascular accident (CVA).** It is the infarction (sudden necrosis) of brain tissue as the result of a loss of blood perfusion stemming from the obstruction or hemorrhage of a cerebral artery. CVA is the third leading cause of death in the United States. CVAs run in families and have a higher incidence in women than in men and in blacks than in whites. Seventy percent of stroke patients are over the age of 65.

Strokes are classified as *thrombotic, embolic,* or *hemorrhagic,* depending on the cause. A **thrombotic stroke** occurs when a thrombus forms and occludes an artery that supplies the brain. The thrombosis itself is usually triggered by arteriosclerosis, but other risk factors include hypertension, smoking, sickle-cell disease, the use of oral contraceptives, arterial inflammation, and dehydration. **Embolic stroke** is usually caused by a fragment of a thrombus that originates outside the CNS, breaks free (becoming a thromboembolus), and travels in the bloodstream until it lodges in a cerebral artery. In some cases, however, the embolus is a traveling air bubble, a mass of agglutinated bacteria, or a bit of fat. Air embolism sometimes develops after surgery, and fat embolism sometimes results from broken long bones that release yellow bone marrow into the circulation.
Hemorrhagic stroke occurs when brain tissue loses its blood supply because of the rupture of a cerebral artery. Hemorrhagic strokes are further classified according to the size of the brain lesion: *petechial* (pinhead-sized), *small* (up to 2 cm in diameter), or *massive* (several centimeters in diameter).

Treatment must begin within 6 hours of occlusion to avoid irreversible brain damage. Drugs are administered to inhibit clotting, restore blood flow, and protect neurons from damage by calcium inflow, free radicals, and excitatory amino acid neurotransmitters. In the case of hemorrhagic stroke, it is important to relieve the increase in intracranial pressure caused by pooled blood (hematoma), sometimes by aspirating the blood. Losses of motor function are treated with physical therapy.

Case Study 20 The Boy Who Didn't Grow

Five-year-old Kyle is brought to a pediatrician for his annual checkup. By his mother's account, Kyle is a normal, healthy child. She is just bringing him in for his childhood vaccinations and to meet his new pediatrician.

During the physical examination, the pediatrician notices that Kyle is small for his age. Kyle's mother tells the pediatrician that although Kyle has always eaten well, he does not seem to grow as quickly as her other children did at that age. When asked about past illnesses, Kyle's mother answers that he has had no major illnesses, but he frequently gets colds and flu. In fact, she says, Kyle seems to have more colds and flu than his brothers and sisters.

With the mother's consent, the pediatrician has his nurse draw a sample of Kyle's blood for analysis. The results of Kyle's CBC are as follows:

Hematocrit (Hct) = 50%

Hemoglobin (Hb) = 14 g/dL

RBC count = $5.1 \times 10^6/\mu L$

WBC count = 8,000/μL

Eosinophil count = 160/μL

Basophil count = 35/μL

Monocyte count = 321/μL

Lymphocyte count = 3,203/μL

Platelet count = 380,000/μL

After studying these test results, the pediatrician reevaluates Kyle's heart and lungs. Respiratory function is normal, but he hears a slight heart murmur. The pediatrician then suggests a chest X ray, which reveals left-ventricular hypertrophy and alterations in the pulmonary blood vessels. After an echocardiogram is conducted, Kyle's condition is diagnosed as patent ductus arteriosus (PDA).

Based on this case study and other information in this chapter, answer the following questions.

1. Does Kyle's CBC reveal any abnormalities in the number, shape, or size of blood cells?
2. Why is a heart murmur heard in both systole and diastole in Kyle's condition?
3. Why does PDA reduce a child's growth rate? Why does it increase the incidence of respiratory infections?
4. Why does the left ventricle sometimes undergo hypertrophy in patients with PDA?
5. If edema were to occur in a PDA patient, would it most likely be localized to the systemic or the pulmonary circulation?
6. Although all aneurysms are serious health concerns, why are dissecting aortic aneurysms more serious than small aortic aneurysms?
7. Nancy has been working on a term paper for the past 3 hours. Upon hearing the doorbell ring, she gets up quickly and faints. Her roommate, Angie, hears Nancy fall and comes running. Assuming that Angie understands the causes of syncope, what could she do to help Nancy?
8. Medical support hose are often used to treat both orthostatic hypotension and varicose veins. How could support hose alleviate the signs and symptoms of either disorder?
9. A patient suffering a hemorrhagic stroke is mistakenly given coumarin. Discuss why you would expect this patient's condition to worsen rather than improve.
10. Do you think orthostatic hypotension could be more successfully treated with a diuretic or an antidiuretic drug? Explain your reasoning.

Selected Clinical Terms

aneurysm A weak point in a blood vessel or in the heart where a thin-walled sac forms and may rupture.

cerebrovascular accident A stroke; infarction of brain tissue due to a loss of cerebral blood flow, resulting from the obstruction or hemorrhage of a cerebral artery.

hemoptysis Spitting up blood from the respiratory tract; usually a sign of tracheal, bronchial, or pulmonary hemorrhage.

orthostatic hypotension A rapid drop in blood pressure that occurs when a person stands up, owing to the gravitational drainage of blood into the lower trunk and limbs without adequate compensation by the baroreflex.

patent ductus arteriosus A congenital failure of the ductus arteriosus (a fetal shunt between the pulmonary trunk and aorta) to close after birth, causing circulatory disturbances that can lead to poor weight gain, respiratory infections, and left heart failure.

Raynaud disease An idiopathic spasm of the arteries of the digits, especially the fingers, in response to conditions such as cold and emotional stress, causing pallor, numbness, coldness, throbbing, and paresthesia.

syncope Fainting; a loss of consciousness and muscle tone, often with falling, due to a loss of cerebral blood flow.

varicose vein A vein swollen with pooled blood, often due to failure of the venous valves to close and prevent backflow of the blood.

21 The Lymphatic and Immune Systems

Objectives

In this chapter we will study

- lymphatic and immune disorders;
- some diseases of the lymph nodes and lymphoid tissues—metastatic cancer, lymphomas, and lymphadenitis;
- tonsillitis;
- infectious mononucleosis;
- autoimmune diseases, particularly systemic lupus erythematosus; and
- how the body's natural immune response affects tissue grafts and transplants.

Diagnosis of Lymphatic and Immune Disorders

The lymphatic and immune systems work together to detect and destroy foreign substances and microorganisms that may disturb homeostasis. In addition, the lymphatic system aids the cardiovascular system in maintaining fluid balance. Disorders of the lymphatic system are therefore often associated with disorders of immunity and fluid balance. Such interactions must be kept in mind in the diagnosis of lymphatic and immune disorders.

Signs of lymphatic system disorders include the following:

- *Fever* indicates infection, while *weakness* and *fatigue* both suggest altered homeostasis.
- *Splenomegaly* (enlargement of the spleen) is common in infections and hematologic disorders. It can be detected by palpation, and an enlarged spleen typically produces a dull sound on percussion.
- **Lymphadenitis** (inflammation of the lymph nodes) usually results from an infection elsewhere in the body, and is marked by enlargement and tenderness of a lymph node. Clinicians palpate the lymph nodes to assess their texture, size, relative mobility, and degree of tenderness. These variables help narrow down the diagnosis.
- **Lymphangitis** (inflammation of the lymphatic vessels) is indicated by the presence of red streaks on the skin (erythematosus), which often radiate from the site of inflammation. Lymphangitis is commonly seen in the lower limbs and was once used to identify "blood poisoning" (now called **septicemia**—bacteria in the bloodstream).
- *Edema* is often an indication of impaired lymphatic drainage of tissue fluid.
- *Respiratory disorders* such as coughing, wheezing, dyspnea, and runny nose are common indicators of immune hypersensitivity (allergy).
- *Skin lesions* (see chapter 7 of this manual) are also seen in response to allergic reactions. The most common type of skin lesion is *hives (urticaria).*
- *Recurrent infections* indicate that the immune system may not be functioning adequately.

The results of a physical examination may indicate the need for additional testing. Possible diagnostic procedures include blood analysis (CBC, immunoglobulin electrophoresis, and identification of specific antibodies), imaging techniques (CT, MRI, and X ray), skin tests to identify allergic reactions, biopsy of the lymphoid tissues, and lymphangiography, a test similar to angiography in which dye is used to visualize the lymph nodes and lymphatic ducts.

Disorders of the Lymphatic System

Although the lymphatic system serves largely to protect us from disease, it sometimes contributes to the spread of disease, and it is also subject to diseases of its own. Here, we consider the role of the lymphatic system in the spread of cancer and examine some diseases of the lymph nodes and tonsils. Lymph node disease in general is called **lymphadenopathy.**

Lymph Nodes and Cancer Metastasis

Because one of the roles of the lymph nodes is to remove foreign substances, the nodes are predisposed to cancer metastasis. When a cancer metastasizes, some of the tumor cells enter the blood and tissue fluid. From the latter site, they are easily picked up by the lymphatic capillaries and transported in the lymph until they lodge in a nearby lymph node. If the cancer cells are not destroyed in the node, they can seed the growth of a metastatic tumor. The degree of metastasis can be determined by finding the lymph nodes most distant from the original site of the cancer that are "normal." This is also the reason the lymph nodes draining the site of a malignant tumor are removed and examined for the presence of abnormal cells at the same time the tumor itself is excised.

Lymphomas

Lymphoma is a collective term for both benign and malignant neoplasms of the lymphoid tissues, although the word is often used alone to mean malignant lymphoma. It is estimated that between 50,000 and 80,000 new cases of lymphoma are diagnosed annually in the United States. Malignant lymphomas are divided into three different types based on cell appearance and origin: Hodgkin disease, non-Hodgkin lymphoma, and Burkitt lymphoma.

Hodgkin disease is a malignant lymphoma first characterized in 1832 by British physician Thomas Hodgkin. It usually affects lymph nodes in the mediastinal, supraclavicular, or cervical region. It occurs almost equally in males and females and is seldom seen before the age of 10. Incidence peaks in people 15 to 34 years of age and again in those over 60. The incidence of Hodgkin disease in the United States in females and males, respectively, is 26 and 35 cases per million. It is apparently caused by an oncogene.

The signs and symptoms of Hodgkin disease include painless swelling of the lymph nodes, splenomegaly, hepatomegaly (liver enlargement), fever, anorexia (loss of appetite), weight loss, night sweats, and pruritis (severe itching). Laboratory analysis reveals thrombocytosis, leukocytosis, eosinophilia, an elevated RBC sedimentation rate, and elevated serum alkaline phosphatase. Diagnosis is made by combining the findings of the physical examination, imaging techniques, and laboratory tests. The disease is confirmed by the presence of characteristic *Reed-Sternberg cells* in a lymph node biopsy. Hodgkin disease is treated with radiation and chemotherapy, with a survival rate of 70% to 80%.

Non-Hodgkin lymphomas are a group of lymphomas in which Reed-Sternberg cells are not observed upon biopsy. Otherwise, the signs and symptoms are similar to those of Hodgkin lymphoma. In non-Hodgkin lymphoma, the lymphadenopathy is not always localized to the cervical and mediastinal lymph nodes, but can include the axillary, inguinal, and femoral lymph nodes. Additionally, these cancers sometimes develop in extranodal sites, including the nasopharynx, bone, thyroid, testes, and gastrointestinal tract.

Non-Hodgkin lymphoma is more common than Hodgkin lymphoma, and its etiology is usually unknown. Patients on immunosuppressive drugs, however, have a 100 times greater risk of developing non-Hodgkin lymphoma than other people. It is thought that the drugs activate a virus which, in turn, causes the genetic transformation to a cancer cell. Diagnosis and treatment methods are similar to those for Hodgkin disease. Monoclonal antibodies against the tumor cells are also effective. Slightly more than half of patients survive, but mortality is higher than for Hodgkin disease. The median time of death ranges from 6 months to 7.5 years after the first symptoms appear, depending on how advanced the lymphoma is when treatment begins.

Burkitt lymphoma affects predominantly children and young adults in central Africa, where it is thought to involve an insect-borne virus. It is rare in the United States. Burkitt lymphoma causes bone-destroying lesions of the jaw and face. It is treated with chemotherapy and is highly curable if treated early, although it disproportionately affects people with little access to the necessary health care.

Lymphadenitis

Lymphadenitis can be triggered by a wide variety of pathogens, including bacteria, viruses, fungi, and protozoans. Streptococcal infections, tuberculosis, cat-scratch disease, primary syphilis, and genital herpes, among other disorders, cause regional lymphadenitis (inflammation of lymph nodes in selected areas), while infectious mononucleosis, cytomegalovirus, secondary syphilis, and other conditions cause generalized lymphadenitis. The lymph node enlargement is a result of edema and infiltration of the node with leukocytes. Lymphadenitis usually subsides when the underlying cause, such as an infection, is cured, but inflamed, abscessed lymph nodes sometimes require drainage.

Tonsillitis

Tonsillitis (inflammation of the tonsils) usually results from streptococcal or viral infections. The pharyngeal and palatine tonsils are most often involved. Signs and symptoms include a sore throat, redness, difficulty swallowing, high fever, headache, malaise, and vomiting. As the tonsils swell with inflammation, they may obstruct breathing. Tonsillitis is most often diagnosed through a physical examination and patient history. A throat culture is often done to determine whether the cause is viral or bacterial and to rule out other disorders. If the tonsillitis has a bacterial cause, an appropriate antibiotic is prescribed; viral tonsillitis is treated with bed rest and aspirin to relieve the pain. Tonsillectomy (removal of the tonsils) is nowadays performed only in the event of frequently recurring tonsillitis.

Infectious Mononucleosis

Infectious mononucleosis is an acute infection of the B lymphocytes with Epstein-Barr virus (EBV). It usually affects young adults between the ages of 15 and 33 years, with peak incidences at the ages of 18 to 23 in males and 15 to 16 in females. At ages younger than these, children in lower socioeconomic groups are most likely to contract EBV, and it is estimated that up to 85% of them do so by the age of 4. However, at this age, children are usually asymptomatic and gain some immunity to further exposure. Most people have acquired an EBV infection by early adulthood, but only a minority of them develop any clinical signs. After the initial infection, the virus remains in the body for life, but is normally kept in check by the immune system.

EBV initially replicates in the nasopharynx, oropharynx, and salivary glands before invading the B lymphocytes. People transmit the virus to others through close contact, usually by exchanging saliva, so mononucleosis is sometimes called the "kissing disease." The incubation period of approximately 30 to 50 days allows time for considerable transmission.

Patients with infectious mononucleosis usually exhibit four signs and symptoms: fatigue, fever, pharyngitis, and lymphadenopathy. The individual may also have malaise, headache, anorexia, and dysphagia. Approximately 50% of all patients have splenomegaly and mild hepatomegaly. Diagnosis is confirmed through blood tests that show leukocytosis, lymphocytosis, and antigens against specific EBV proteins.

Mononucleosis is usually self-limiting, lasting only a few weeks after diagnosis. Treatment is therefore primarily supportive and includes bed rest, analgesics for pain, and antipyretics for fever. In rare cases, death occurs from splenic rupture or airway obstruction. Strenuous physical activity should be restricted for about 2 months to avoid the last of these complications.

Autoimmune Diseases

Autoimmune diseases are disorders in which the immune system mistakenly recognizes normal tissues as foreign and initiates immune-mediated destruction of the tissue cells. Virtually every body system is affected by autoimmune diseases, some of which have been discussed in previous chapters of this manual (for example, rheumatoid arthritis in chapter 10; myasthenia gravis, chapter 12; and Graves disease, chapter 17). Most autoimmune disorders appear to involve a genetic predisposition. They affect more women than men.

Systemic lupus erythematosus (SLE), is one of the most serious autoimmune diseases. Approximately 90% of SLE patients are women between the ages of 20 and 40; blacks are affected more frequently than whites. In SLE, autoantibodies are produced against a wide variety of substances, including blood cells (erythrocytes, platelets, and lymphocytes), clotting proteins, phospholipids, and especially nuclear contents such as the nucleic acids and histones. Thus, virtually every cell of the body is subject to attack. Tissue damage most often occurs when antibody-antigen complexes are deposited in body tissues. The most common site of deposition is the basement membrane of the glomerulus in –the kidney, leading to renal complications.

Manifestations of SLE include arthralgia (joint pain) or arthritis, vasculitis (inflamed blood vessels), rash, renal and cardiovascular dysfunction, and anemia or other blood disorders. Diagnosis is complicated by the fact that there are periods of remission during which the patient is asymptomatic.

There are 11 clinical indicators of SLE; a patient must present with at least four of these to be diagnosed with the disease:

1. malar rash (a rash confined to the cheeks);
2. discoid rash (showing raised, scaly patches);
3. photosensitivity (a rash triggered by sunlight);
4. oral or nasopharyngeal ulcers;
5. arthritis in at least two joints;
6. inflammation of the serous membranes;
7. renal dysfunction (often with protein in the urine);
8. neurologic dysfunctions such as seizures or psychosis;
9. hematologic disorders such as anemia, leukopenia, or thrombocytopenia;
10. presence of antibody against nuclear contents; and
11. various other abnormal antibodies and serological signs.

Treatment and prognosis for SLE depend on the severity of the symptoms and the systems that are affected. In most countries, the 10-year survival rate exceeds 95% if diagnosis is made promptly. Treatment is aimed at managing the signs and symptoms, and includes steroidal anti-inflammatory and analgesic medications.

Transplant Rejection

As with autoimmune diseases, clinicians might wish for a less active immune system when transplanting tissues and organs into a patient because a normal, healthy immune system does its best to destroy transplanted foreign tissues.

Transplant rejection is an example of *alloimmunity*—immune responses to cells that are genetically different from the host body but belong to the same species. To minimize this reaction, it is necessary to have a tissue donor who is antigenically compatible with the recipient. Tissue compatibility is determined by **HLAs** (an abbreviation for either **human leukocyte antigens** or **histocompatibility locus antigens**), coded for by genes called the *major histocompatibility complex (MHC)*. A perfect HLA match is possible only between identical twins; other siblings have just a 1 in 4 chance of being antigenically compatible. The greater the difference in HLAs, the greater is the probability that the transplant will be rejected. Finding a compatible donor is difficult and important.

When the transplant recipient's immune system detects the nonself antigens on the transplanted tissue, immune responses are triggered. These rejection responses are characterized as hyperacute, acute, or chronic, depending on the time course. *Hyperacute rejection* is rare, but occurs almost immediately after blood perfusion to the transplanted organ is established. It occurs in recipients who already have antibodies against antigens of the transplanted tissue. *Acute rejection* occurs about 2 weeks after the transplant as the recipient develops antibodies against the donor's HLA antigens. *Chronic rejection* occurs after months to years of normal transplant function as a result of a weak immune response against minor HLA antigens in the transplant. It results in slow, progressive organ failure.

With improved medical technology, including advances in surgical techniques and the development of immunosuppressive drugs, the number of successful transplants has increased. Immunosuppressive drugs allow for transplants between less compatible donors and recipients, although a certain degree of compatibility is still required. By suppressing the immune system, these drugs lessen the likelihood of rejection and give the transplanted tissue time to become established. However, immunosuppressive drugs also make the patient more susceptible to opportunistic infections, and it may be months before normal immune function returns. To minimize the development of infections, patients on immunosuppressive drugs are often treated with immunoglobulins (gamma globulins) and antibiotics.

Case Study 21 The Graduate Student with Swollen Lymph Nodes

Zach is a 28-year-old graduate student working in a laboratory that studies the genetics of the human immunodeficiency virus (HIV). He has recently noticed that he is losing weight, his lymph nodes are swollen, and he has been experiencing night sweats. He also seems to be scratching more often than normal. Zach is concerned that he may have contracted HIV through his research project.

Zach is married, monogamous, and has never received a blood transfusion or used intravenous drugs. Physical examination reveals pallor, lymphadenopathy, splenomegaly, and an abnormal mass in his abdomen. Zach's heart and respiratory sounds are normal, but his body temperature is slightly elevated (99°F). Blood analysis is done, with the following results.

Hematocrit (Hct) = 35%

Hemoglobin (Hb) = 9.5 g/dL

RBC count = $3.5 \times 10^6/\mu L$

WBC count = 22,000/μL

Platelet count = 450,000/μL

Alkaline phosphatase = 120 IU/L

HIV antibodies = Negative

IgA = 300 mg/dL

IgG = 1,500 mg/dL

IgM = 65 mg/dL

Erythrocyte sedimentation rate = 23 mm/hr

Differential WBC count:

Neutrophils = 65%

Eosinophils = <1%

Basophils = 1%

Lymphocytes = 30%

Monocytes = 3%

Suspecting a lymphoma, the physician suggests a lymph node biopsy. Results of the biopsy show fibrosis, few lymphocytes, and the presence of Reed-Sternberg cells. These results confirm a diagnosis of Hodgkin disease.

Based on this case study and other information in this chapter, answer the following questions.

1. Other than his occupation, what symptoms lead Zach to suspect HIV infection?
2. What signs, symptoms, and test results lead his physician to suspect a lymphoma?
3. Why is Zach diagnosed specifically with the Hodgkin form of lymphoma?
4. What treatment will likely be prescribed for Zach? What is his prognosis?
5. Marcy has been diagnosed with renal failure and is admitted to the hospital for a kidney transplant. What postsurgical treatment should be initiated to prevent rejection of the organ?
6. What complication of the postsurgical treatment in the previous question would be of greatest concern to Marcy?
7. Why is infectious mononucleosis more prevalent in adolescents and young adults than in other age groups?
8. Suppose a man had an autoimmune disease in which his body produced autoantibodies against sperm cells. What would you expect to be the chief complaint?
9. Sandra presents to her physician complaining of fever, headache, and painful swellings in her armpits. During the physical examination, the doctor notices that her hands are covered with scratches. He asks about these, and Sandra says they're from a kitten that she found wandering along a country road 2 weeks earlier and adopted as a pet. What disorder do you think her physician might suspect?
10. Why are antibiotics prescribed for bacterial tonsillitis but not for viral tonsillitis?

Selected Clinical Terms

infectious mononucleosis An acute infection of the B lymphocytes with the Epstein-Barr virus, found mostly in adolescents and young adults and typically causing a few weeks of malaise, headache, anorexia, and dysphagia.

lymphadenitis Inflammation of a lymph node, marked by swelling and tenderness; usually indicative of an infection in a region of the body whose lymphatic drainage leads to that node.

lymphadenopathy A collective term for all diseases of the lymph nodes.

lymphangitis Inflammation of a lymphatic vessel.

lymphoma Any neoplasm of the lymphoid tissues, especially malignant neoplasms.

septicemia The presence of bacteria in the bloodstream; formerly called blood poisoning.

systemic lupus erythematosus An autoimmune disease that involves widespread immune attack on the body's tissues, producing renal complications, connective tissue disease, a characteristic facial rash, and other pathologies.

tonsillitis Inflammation of the tonsils.

22 The Respiratory System

Objectives

In this chapter we will study

- methods used to diagnose respiratory disorders;
- sudden infant death syndrome;
- tuberculosis;
- some physiological effects of high altitude, especially mountain sickness;
- cystic fibrosis;
- asthma; and
- an occupational lung disease, pneumoconiosis.

Diagnosis of Respiratory Disorders

The most common symptoms of respiratory disease are chest pain, cough, and dyspnea. The pain usually worsens when the patient breathes or coughs. Coughing is a complex reflex that serves to clear the airway of irritants. It helps protect the lungs against infection and aspiration of fluid. A **productive cough** is one that brings up **sputum,** a thick liquid composed of mucus, cellular debris, bacteria, and sometimes blood or pus. A **nonproductive cough** is dry. The onset, duration, and frequency of coughing can help identify a specific respiratory disorder. Dyspnea is a symptom of many pulmonary disorders, but it can also be a sign of cardiovascular or metabolic problems. When making a diagnosis, the clinician checks for any variation in the degree of dyspnea and whether exertion induces or worsens it.

Other signs and symptoms are also characteristic of respiratory disorders. For example, **clubbing** of the fingers or toes (bulbous enlargement of the distal segment of the digits) is common in diseases that result in decreased oxygen delivery to the tissues. Cyanosis (bluish discoloration of the skin and mucous membranes) is produced in response to reduced hemoglobin concentrations or increased amounts of deoxygenated hemoglobin. Hemoptysis (coughing up blood or bloody sputum) often indicates damage to the lung tissue or the bronchi. Blood produced in sputum is normally bright red with an alkaline pH, while blood that is vomited is darker (resembling coffee grounds) and has an acidic pH. Sputum can be assessed for color, consistency, appearance, odor, and amount produced. It may also be cultured to identify the microorganism causing a disease.

During the physical examination, the clinician observes the chest cavity for abnormal dimensions that develop in certain respiratory disorders—for example, people with emphysema often develop "barrel chest." The clinician also palpates the bones and muscles of the thoracic cage for structural abnormalities or asymmetry and uses a stethoscope to listen to the breathing. Various respiratory disorders are characterized by abnormal respiratory sounds, such as **rales, stridor,** or **wheezing,** or friction rub due to pleural membrane abnormalities. Also during the physical examination, the rate and depth of respiration are evaluated. Even in the absence of dyspnea, changes in respiratory rate and depth suggest respiratory disorders.

The information obtained from the patient history and physical examination determines which imaging or laboratory tests should be conducted and whether *spirometry* is needed. Imaging techniques include X ray, CT, and MRI as well as bronchoscopy and thoracoscopy. An endoscope is used to view the larynx, trachea, and bronchial tree in **bronchoscopy** and to view the pleural cavity in **thoracoscopy.** Laboratory tests include sputum analysis and culture, lung biopsy, measurement of arterial blood gases, and analysis of pleural fluid.

Sudden Infant Death Syndrome

Sudden infant death syndrome (SIDS), also known as *crib death,* denotes any sudden, unexpected death of an infant for which no cause can be identified on autopsy. Up to 10,000 infants die of SIDS each year, at a rate of 1.5 to 2.0 per 1,000 live births. SIDS is the leading cause of death in infants between the ages of 1 month and 1 year. Most of these deaths occur between 2 and 4 months of age. Death most frequently occurs between midnight and 6 a.m., during the winter months, and almost always while the infant is asleep, usually in a prone position. The seasonal variation suggests that an environmental factor such as a virus may play a role in SIDS.

The risk factors for SIDS include low birth weight, premature delivery, smoking during pregnancy, anemia during pregnancy, lack of prenatal care, and siblings who died of SIDS. Some studies suggest that SIDS results from pulmonary edema that is triggered by neutrophil degranulation and leads to respiratory obstruction, hypoxia, and death.

To minimize crib deaths, recommendations include placing at-risk infants on apnea monitors, training parents and caregivers in infant cardiopulmonary resuscitation, and maintaining a cool ambient temperature. It is also recommended that infants never be put to bed face-down.

Tuberculosis

Tuberculosis (TB) is caused by the bacillus *Mycobacterium tuberculosis*. Before the development of effective antibiotics, TB was a leading cause of death worldwide. It was long known as *consumption,* because of the way patients wasted away; the name *tuberculosis* came into use after 1860. Although the incidence of tuberculosis in North America decreased between 1950 and 1980, it increased again after 1985, especially in men between the ages of 25 and 44. One reason for this increased incidence is AIDS, which leaves its immunocompromised victims unable to ward off respiratory infections. Another reason is an increase in the number of antibiotic-resistant strains of tuberculosis bacteria. This is the result of the failure of patients to take medications consistently, allowing drug-resistant bacteria to survive and multiply. Other contributing factors are emigration, homelessness, substance abuse, institutional living arrangements, and lack of access to medical care.

Tuberculosis is transmitted from one person to another by airborne droplets. Once infection occurs, the bacteria migrate to the lungs where they are engulfed by neutrophils and alveolar macrophages. These phagocytes surround and isolate bacterial colonies, forming small fibrous lesions called *tubercles.* Over a course of about 10 days, the center of the tubercle undergoes necrosis, and scar tissue grows around the tubercle. Bacteria can remain dormant in these tubercles for many years, but can be reactivated later in life and spread through the blood and lymphatic system to other organs.

Signs and symptoms of tuberculosis include fatigue, weight loss, lethargy, anorexia, and low-grade fever. The patient may have night sweats and a cough that produces **purulent** sputum (sputum with pus) or blood. Diagnosis is by chest X ray, sputum culture, and a positive tuberculin skin test. However, it should be noted that a positive skin test alone is not a sure indicator of TB because some individuals develop antibodies through exposure but do not have the disease. Sputum culture allows isolation of *M. tuberculosis*. A chest X ray shows nodules and other changes characteristic of the disease.

Treatment employs antibiotics to control the bacterium and prevent transmission. The drug selected is based on the bacterial strain isolated, the health of the individual, and the presence or absence of active disease. Patients with active tuberculosis are isolated at home or in the hospital until sputum cultures are free of active bacteria. Depending upon the drugs selected, the patient may be in isolation for as long as 2 months. The chance for a full recovery is excellent if the patient completes the drug treatment.

Mountain (Altitude) Sickness

The respiratory function of humans and other animals is adapted to the air pressure, and therefore the altitude, at which they live. People can adapt to gradual changes in pressure, but sudden, extreme changes cause various cardiovascular and pulmonary dysfunctions.

Ascending quickly to high altitudes, as in aviation and mountain climbing, exposes people to low atmospheric pressures and causes another set of problems. About 20% of those who ascend to 2,700 m (9,000 ft) in less than 1 day develop signs and symptoms of **mountain (altitude) sickness.** Within 24 hours at the new altitude, a person begins to experience such symptoms as insomnia, tachypnea, headache, nausea, and vomiting, which worsen upon exertion. These problems subside in a few days as the body acclimates to the altitude. Acclimation includes increases in hematocrit and thus the oxygen-carrying capacity of the blood, increased capillary density in the muscles and other tissues, and more myoglobin and mitochondria in the muscles.

The pathogenesis of mountain sickness is not entirely clear. It stems ultimately from the low oxygen partial pressure at high altitude. Hypoxia increases the respiratory rate. This response helps to better oxygenate the tissues, but it "blows off" carbon dioxide faster than the body produces it. Thus, it induces a state of respiratory alkalosis. In addition, the Na^+-K^+ pumps of all the body's cells work less well in conditions of hypoxia, so Na^+ and water accumulate in the cells and the cells become swollen. Pulmonary edema, cerebral edema, and cerebral hemorrhages are often found in autopsies of people who die from mountain sickness, and the blood vessels are often congested with thrombi and emboli. Cerebral edema can cause mental confusion, hallucinations, and a loss of motor coordination. Retinal hemorrhages are common above 5,000 m, but usually cause no symptoms and clear up after return to lower altitude. Complete blindness can occur at very high altitudes, however.

Mountain sickness is best avoided by ascending slowly—2 days for the first 2,500 m (8,000 ft) and 1 day for every 600 m (2,000 ft) above that—and drinking ample water. Physical fitness provides no protection against mountain sickness. Little treatment is required except for rest, a light diet, fluid, analgesics, and sometimes descent.

Cystic Fibrosis

Cystic fibrosis (CF) is an inherited disorder that affects primarily the respiratory and digestive tracts. It is the most common lethal hereditary disease of whites, affecting about 1 in every 3,500 white children and 1 in 12,000 black children, but few of Asian descent. The most common cause of death is respiratory failure.

The gene responsible for CF has been mapped to chromosome 7 and codes for a chloride ion transport protein, CFTR. Mutated forms of CFTR do not properly transport chloride ion (Cl^-). As a result, mucous cells produce an abnormally thick, sticky mucus that obstructs the airway, pancreatic ducts, and some parts of the reproductive tract. Respiratory cilia are virtually immobilized by this mucus, so the mucus remains stationary and provides a breeding ground for chronic respiratory infections. Excess NaCl appears in sweat and other serous secretions; unusual saltiness of a baby's sweat can be one of the first signs of CF. Cystic fibrosis occurs only in those who are homozygous recessive for the CFTR gene, although heterozygous carriers may have asymptomatic Cl^- transport abnormalities. Specific genetic markers for CF have been identified, allowing prenatal diagnosis and genetic counseling.

Children with CF tend to grow poorly because mucus obstructing the pancreatic ducts interferes with the secretion of digestive enzymes. Dietary therapy, however, can control the effects of CF on digestion, leaving respiratory dysfunction as the most conspicuous and life-threatening effect. The mucus produced by the goblet cells of CF patients collects in the pulmonary air passages, blocking and dilating them. This predisposes the lungs to chronic infections with such bacteria as *Staphylococcus aureus* and *Pseudomonas aeruginosa.* The pulmonary syndrome progresses to chronic bronchitis, pneumonia, pulmonary fibrosis, atelectasis, alveolar hemorrhaging, and cor pulmonale. Hypoxia and respiratory obstruction lead to clubbing of the digits, cyanosis, and a barrel chest.

One way to diagnose CF is through a sweat test. In this procedure, sweating is induced with pilocarpine, and the electrolyte concentrations of the sweat are measured. Concentrations of Na^+ or Cl^- greater than 60 mEq/L confirm CF. Genetic testing for CF has now been developed and, if available through the clinical facility, may be used instead of or in addition to a sweat test.

Presently there is no cure for CF, so the treatment goal is to minimize complications and enable the patient to live as normal a life as possible. Following diagnosis, the CF patient is placed on a specific diet to minimize the gastrointestinal complications. Also, because the pancreas is affected, the patient must be supplied with the enzymes the pancreas would normally secrete. Prompt antibiotic treatment of respiratory infections minimizes complications. In addition, chest physical therapy is used, including percussion (striking the chest to help clear the mucus), assisted coughing, and placing the patient in a posture that facilitates chest drainage. If the patient is in severe respiratory distress, oxygen therapy may be used. **Mucolytics** and **expectorants** are used to degrade mucus and to promote its elimination from the respiratory tract, respectively.

Surgery is sometimes required for patients who develop further complications, such as pneumothorax, gallbladder disease, or chronic sinusitis. Studies are currently in progress to evaluate the feasibility of using gene therapy to treat CF. In gene therapy, a "normal" copy of the CFTR gene would be introduced into the cells of the respiratory system, with the goal of successfully restoring the function of the CFTR. Minimizing the respiratory complications in this way could lead to a longer life for CF patients.

Despite some advances, the prognosis for CF is still poor. Individual life spans vary greatly from patient to patient, but the average life expectancy is 30 years. Death usually results from complications of respiratory infection. Long-term survival is greatest in males, blacks, and patients with minimal pancreatic or gastrointestinal involvement.

Asthma

Asthma is one of three major chronic obstructive pulmonary diseases (COPDs). The other two—chronic bronchitis and emphysema—are primarily consequences of smoking. Asthma, however, is a hereditary immune disorder. It results from a complex combination of hereditary and environmental risk factors. Asthmatics suffer chronic inflammation and hypersensitivity of the airway. Environmental antigens, irritants, and even emotional states can trigger **bronchospasm,** a prolonged contraction of the bronchioles that results in labored breathing *(dyspnea)* and sometimes suffocation. Relief can be obtained from bronchodilators such as epinephrine, taken by means of an inhaler. Nevertheless, about 5,000 people die of asthmatic attacks annually in the United States.

Asthma takes multiple forms, and specialists are not entirely settled on the best way to classify them. A traditional system, with some admitted shortcomings, is to distinguish **extrinsic asthma** from **intrinsic asthma.** Extrinsic asthma is an allergic response to environmental antigens that come from such household pests as cockroaches, dust mites, pollen, animal dander, and molds. Intrinsic asthma involves no identifiable allergy, and is triggered by such factors as cold air, aspirin and other drugs, exhaust fumes and other air pollutants, physical exertion, and emotion. Extrinsic asthma is the more common form and disproportionately affects children. Two-thirds of cases appear before age 40. Intrinsic asthma usually affects adults over 35.

Asthma is the most common chronic illness of childhood, and its incidence is increasing at an alarming rate. Since 1980, the number of children with asthma has doubled in the United States, where it is associated especially with poor sanitation and ventilation, crowding, lack of exercise, and staying indoors most of the time. Yet, outside of the U.S., it is often associated with extreme cleanliness, leading to a theory that modern hygiene keeps some environments so clean that the childhood immune system is under-challenged and does not develop normal responsiveness. Asthma is most common in countries where vaccines and antibiotics are widely used. It is relatively uncommon in developing countries and among farm families, where children are heavily exposed to environmental antigens and develop healthy, responsive immune systems.

Patient education and the correct use of inhalers can reduce the number, duration, and severity of asthmatic attacks as well as the need for hospitalization. However, the preventive treatment regimen can involve taking as many as eight different medications per day, and patient compliance remains a major obstacle to treatment.

Pneumoconiosis

Many lung diseases are caused by inhaling substances such as silica dust, coal dust, asbestos dust, irritating gases, allergens, and carcinogens, usually while on the job. The physiological effects vary, depending on the substances inhaled and the individuals exposed. For example, particulate matter is normally trapped and removed by the mucociliary escalator, but this protective mechanism is compromised in smokers, who are therefore more susceptible to particulate matter than their nonsmoking coworkers.

Pneumoconiosis is a family of restrictive lung diseases caused by long-term inhalation of inorganic dust particles, typically in the workplace. Pneumoconiosis usually begins after years of exposure to particulate matter and produces some signs and symptoms similar to those seen in a long-term smoker. The most common causes of pneumoconiosis are silica, asbestos, and coal dusts, but other dusts can also induce it (talc, fiberglass, cement, and others). Regardless of the type of dust, the deposition of particles in the lungs is permanent, causing pulmonary fibrosis.

Diagnosis is based on the history of exposure, physical examination, chest X ray, sputum analysis, and spirometry. Pneumoconiosis is identified clinically according to the dust particle that causes it. Probably the most widely known form of pneumoconiosis is *black lung disease,* caused by coal dust. Silica and asbestos produce types of pneumoconiosis called *silicosis* and *asbestosis,* respectively. Beginning in the 1940s, asbestos was widely used in buildings for electrical and thermal insulation, ceiling and floor tiles, and other purposes. Its use was phased out beginning in 1975 because of the recognition that it not only causes pneumoconiosis, but also greatly increases the risk of squamous cell carcinoma of the lung and *mesothelioma,* a cancer of the pleurae that is almost invariably fatal.

The signs and symptoms of the various forms of pneumoconiosis are very similar; they include cough, dyspnea, wheezing, and exercise intolerance. Spirometry reveals reduced lung volumes. Analysis of sputum and lung biopsy are sometimes necessary to rule out infection, cancer, and other possible causes of similar symptoms. A CBC may reveal eosinophilia and polycythemia.

Treatment has two primary goals: to palliate the symptoms and to minimize further exposure. Unfortunately, particle exposure is often unavoidable in the patient's workplace, and finding a different job may not be feasible. However, the symptoms can be lessened by managing infection, using bronchodilators to minimize bronchospasm and inflammation, and using oxygen therapy when necessary. Also, if the patient smokes, he or she is encouraged to stop.

Case Study 22 The Teacher with a Persistent Cough

Nick is a 58-year-old schoolteacher who sees his physician with a respiratory complaint. He says that for the last 2 years he has had a persistent dry cough that seems to be getting worse, sometimes interfering with his sleep or interrupting his classroom lectures. He also says that he seems increasingly unable to tolerate heavy work. He reports that he recently became seriously out of breath just changing a tire on his pickup truck, and in ordinary tasks such as raking the lawn, he has to stop frequently and catch his breath. Occasionally, he even finds himself gasping for breath when simply sitting and watching television.

His doctor notes a slight cyanosis of Nick's fingernail beds and clubbing around the edges of the nails. Upon auscultation, he hears a dry inspiratory crackling in the lower lobes of the lungs. He refers Nick to the hospital radiology department for a chest X ray and to a pulmonologist for some pulmonary function tests. A nurse draws blood for hematology.

The hematologic results show a normal blood pH and PCO_2, but moderate hypoxemia. The pulmonary function tests show that Nick's FEV_1 is normal, but he has a significantly reduced vital capacity (VC) and total lung capacity (TLC). The chest X ray exhibits calcified plaques on the parietal pleura, long opaque streaks in the lower lobes of the lungs, and curved opaque lines about 5 to 10 cm long parallel to the pleural surface. Parts of the lung are starting to look honeycombed.

The doctor calls Nick in to discuss the findings and interviews him in great depth on his employment history. Nick says that after high school, he worked from 1958 to 1970 on a building demolition crew before deciding to go to college and get a teaching certificate. After completing his master's degree, he began teaching in 1976 at a school that had just been built, and has taught there for 22 years. Nick says that his mother, a heavy smoker, died of lung cancer at age 64, but he knows of no other respiratory disease in the family. He says that he smokes about a pack and a half of cigarettes a day.

In light of the clinical findings and Nick's occupational history, the doctor diagnoses the problem as asbestosis. He shows Nick the chest X ray and says the lesions are very consistent with this disease, which would also account for Nick's symptoms and the findings on his blood gases and pulmonary function tests. Nick is surprised at this because he knows of no asbestos exposure, but the doctor says he could have inhaled asbestos dust during the years he worked tearing down old buildings. "But that was a long time ago," Nick protests. The doctor explains that the symptoms of asbestosis often don't appear until many years after exposure. He says nothing can be done to reverse the damage, but he can prescribe an antitussive to ease the coughing attacks and a bronchodilator to improve breathing. He also recommends that Nick get annual flu shots because he has an increased risk of influenza. Most of all, he strongly urges Nick to stop smoking, explaining that asbestosis sharply increases the risk of lung and pleural cancer in smokers.

Based on this case study and other information in this chapter, answer the following questions.

1. What signs and risk factors does Nick present with that suggest a pulmonary disorder?
2. What aspects of Nick's physical examination, chest X ray, and spirometry confirm this hypothesis?
3. Asbestosis, like bronchitis, can be treated with inhalers. Based on your understanding of the autonomic nervous system, what type of receptor agonists or antagonists could be used in an inhaler for this purpose?
4. What signs does Nick have in common with children who have cystic fibrosis? What aspect of his hematologic results is consistent with these signs?
5. What is the relevance of the dates when Nick worked in building demolition? Why is the age of the school where he works now relevant? What might the doctor have done differently if the school had been built in 1960?
6. Marcus, a Miami college student, goes on a skiing trip to Denver for the Christmas holiday. Upon arrival, he quickly joins the other skiers on the slopes. After two runs, he feels nauseated and cannot catch his breath, but he rides the chair lift up for another run anyway. On the way up, he begins vomiting. Marcus thinks he is coming down with the flu, but a ski instructor who works at the resort says he more likely has something else. What would be your guess?
   a. pneumonia
   b. pneumoconiosis
   c. chronic bronchitis
   d. mountain sickness
   e. asthma
7. Patients hospitalized for tuberculosis are often required to wear masks when leaving their rooms, and their visitors must wear masks while in the room. What is the reason for this?
8. Why do you think blood that is vomited is darker than blood that is spit up in hemoptysis?
9. What is the purpose of culturing sputum in the diagnosis of respiratory disorders? Name a disorder in which this would be especially useful.
10. Tuberculosis is especially prevalent in dark, poorly ventilated homes and institutions where people are crowded and spend most of their time indoors. Explain why.

Selected Clinical Terms

bronchoscopy Viewing the larynx, trachea, and bronchial tree with an endoscope (bronchoscope).

bronchospasm Prolonged constriction of the bronchioles due to contraction of the smooth muscle of the wall, resulting in coughing, wheezing, or dyspnea.

clubbing Growth of the distal segments of the fingers and toes, causing them to become widened and thickened, with abnormally curved, shiny nails; a sign of chronic hypoxemia and various other pathologies.

expectorant An agent that promotes secretion by the respiratory mucosa and promotes the expulsion of secretions from the airway.

mucolytic An agent that dissolves or liquifies mucus and thus loosens up respiratory congestion.

purulent Exhibiting or consisting of pus, or characterized by its formation.

rales An abnormal respiratory sound heard by auscultation of the chest, having either a musical pitch or a crackling sound (the word is used in different senses by different authorities); a sign of bronchospasm or congestion.

sputum Material expelled from the respiratory tract by coughing or clearing the throat; composed of mucus, cellular debris, bacteria, and sometimes blood or pus.

stridor Noisy, high-pitched respiratory sounds, like blowing wind, heard by auscultation of the chest; a sign of respiratory obstruction, especially in the trachea or larynx.

thoracoscopy Viewing the pleural cavity with an endoscope.

wheezing Whistling sounds upon inspiration and expiration.

23 The Urinary System

Objectives

In this chapter we will study

- the common signs of urinary system disorders;
- diagnostic procedures applicable to urinary system disorders;
- urinary tract infections, particularly cystitis and pyelonephritis;
- glomerulopathies such as glomerulonephritis and nephrotic syndrome;
- neurogenic bladder; and
- how diuretics can be used to treat certain urinary disorders.

Assessment of the Urinary System

The urinary system carries out a wide variety of homeostatic functions. It regulates acid-base balance, body water (and blood pressure), and mineral balance. It also plays a vital role in removing metabolic waste products and detoxifying and removing chemicals from the body. Disorders of the urinary system therefore have a far-reaching impact on the other body systems.

Common Signs and Symptoms of Urinary Disorders

As with other body systems discussed previously, some of the signs and symptoms of urinary system disorders—such as fever, malaise, and weight loss—are nonspecific. Therefore, the patient history, physical examination, and clinical tests are essential to achieving an accurate diagnosis.

Specific symptoms of urinary problems include changes in urinary frequency, abnormal urination patterns, changes in the volume of urine produced, abnormal appearance of the urine, and pain upon urination. In addition, individuals with urinary system diseases often have generalized edema and hypertension because their kidneys are unable to properly regulate the body's water and sodium ion balance.

Changes in Urinary Frequency The average adult voids 700 to 2,000 mL of urine over the course of a day, usually distributed in four to six voids. If the frequency of urination increases without a corresponding increase in volume, a number of pathologies are suggested. These include urinary bladder trauma or infection (**cystitis**), urethral infection (**urethritis**), prostate enlargement, production of acidic urine, tumors in the urinary tract, and damage to either the urinary system or the regions of the nervous system that regulate it.

Abnormal Urination Patterns **Urinary incontinence** (**enuresis,** the inability to retain urine) may be a "normal" effect of aging in both men and women, or it may be a sign of urinary system pathology. In women, stretching of the pelvic floor during childbirth may result in incontinence associated with mild physical stress such as sneezing or coughing **(stress incontinence).** Other causes of incontinence include damage to the urinary tract due to childbirth or prostate removal, neurogenic bladder dysfunction, or a congenital defect such as *exstrophy* of the bladder, in which the bladder is everted due to the absence of portions of the lower abdominal wall and anterior wall of the bladder. Incontinence may be prevented in part by doing Kegel exercises—rhythmically tightening the pelvic muscles as if trying to stop the urine stream. This strengthens the muscles of the pelvic floor and reduces the incidence of incontinence.

During the first 2 to 3 years of life, bedwetting **(nocturnal enuresis)** is relatively common. But after 3 years of age, continued enuresis may indicate neurogenic bladder disease, impaired or delayed neuromuscular development of the urinary tract, or a behavioral problem. **Nocturia** (having to void during the night) may simply be due to excessive fluid intake in the late evening—or it may indicate the presence of renal or prostate disease.

Changes in the Volume of Urine Produced Changes in daily urine volume may signal disease.

Polyuria, or *diuresis* (daily urine volume exceeding 2 L/day), suggests that the ability to concentrate urine has been impaired or that the kidney has been damaged. In **oliguria** (urine production of less than 500 mL per day) or **anuria** (urine production of less than 100 mL per day), the patient is at risk of *azotemia* (the accumulation of nitrogenous wastes in the blood) or *uremia* (toxic effects of the accumulated wastes). Oliguria and anuria can be caused by an obstruction of the urinary tract, renal ischemia, congestive heart failure, shock, or pyelonephritis.

Abnormal Appearance of the Urine Changes in the appearance (and odor) of the urine may simply be due to something the patient has eaten, or they may indicate a pathology. The color of urine can vary considerably, ranging from a deep amber to a light, almost colorless yellow. These color variations reflect the body's state of hydration. In normal urine samples, color intensity is an indicator of urine concentration. The darker the color, the more concentrated the urine. Food pigments excreted in the urine are usually red or yellow, while excretion of B vitamins causes the color to become a brighter yellow. Additionally, certain drugs excreted in the urine can cause the color to change from yellow to brown, black, blue, green, or red. If diet and drugs can be ruled out as causes of a color change, disease may be indicated. For example, myoglobin, hemoglobin, and erythrocytes in the urine produce a red or brown color, while bilirubin causes the urine to appear bluish-green to brownish-black. A white, cloudy appearance can occur if the urine contains pus, bacteria, lipids, or semen, or if the urine is alkaline and salts precipitate from solution. Frothy or foamy urine indicates excessive amounts of protein. The exact cause of a change in the appearance of the urine can be determined through urinalysis.

Pain upon Urination Pain or discomfort during urination **(dysuria)** is most often localized to the site of trauma or infection. It can be an extremely important diagnostic sign. Dysuria is a common symptom of cystitis or urethritis or of a urinary tract obstruction such as renal calculi (kidney stones, composed of crystallized urinary salts) or an enlarged prostate gland. When the pain is localized to the lumbar back and radiates laterally and to the upper quadrant (right or left), kidney infection or trauma is the most likely cause. Pain superior to the pubic region suggests cystitis.

Tests Used in Diagnosing Urinary System Disorders

If a clinician suspects a urinary system disorder based on the physical examination and patient history, a number of diagnostic procedures are available to determine whether the system is functioning normally or not. The most commonly used noninvasive test is **urinalysis,** in which a freshly voided urine sample is analyzed. Urinalysis is also used to screen for drug use (legal or illegal) because of the kidney's important role in removing metabolic wastes and toxins. This procedure is becoming more widely used by businesses and the governing bodies of various sports (for example, the Olympic Games and national sports teams) to ensure compliance with drug use policies.

Urinalysis provides a wealth of diagnostic data. Chemical techniques are used to determine the amount of protein, glucose, ketones, blood, nitrites, and hydrogen ions (pH) in the urine as well as whether metabolites of drugs or toxins are present. Urinalysis has been made easier by the development of test strips that are dipped into the urine. Each strip has chemical reagents that change color in the presence of various urine solutes.

Elevated amounts of one or more chemicals may indicate various disorders of the urinary and other systems. For example, *proteinuria* (protein in the urine) is a common indicator of damage to the glomeruli. *Ketonuria* (ketones in the urine) suggests a condition causing metabolic acidosis such as diabetes mellitus or starvation. *Glycosuria* (glucose in the urine) is diagnostic of various forms of diabetes, usually diabetes mellitus.

The solute concentration of the urine can be measured with an *osmometer*, while a *urinometer* may be used to determine the *specific gravity* of a urine sample. Changes in osmolarity or specific gravity suggest either increased or decreased numbers of particles (ions, molecules, and cells) in the urine. Coupled with other tests, a change in urine osmolarity or specific gravity may support the diagnosis of a disease.

In addition, urine can be centrifuged to concentrate its solids, and the sediment examined microscopically for the presence of cells or **casts** (cylindrical masses of mucoproteins formed in the kidney tubule that may contain cells, protein, or fat). The significance of casts in some urinary diseases is explained later in this chapter. Some epithelial cells are expected to be present in the urine due to the natural shedding of cells by the urinary tract. But cells that shouldn't be present in large numbers include blood cells (erythrocytes or leukocytes) and bacteria. Erythrocytes suggest inflammation, tumors, renal calculi, or trauma; leukocytes indicate a urinary tract infection (UTI), and bacteria, of course, confirm this. Urine cytology (examining the cells in urine) sometimes reveals urinary tract cancer.

In addition to urinalysis, many renal function tests have been developed to identify kidney diseases and measure their severity. These tests usually involve measuring the clearance of a specific chemical from the body; this is determined by comparing the concentration of the chemical in blood and urine samples. The clearance of a specific compound can be used to calculate the renal plasma flow or glomerular filtration rate.

The ability of the kidneys to concentrate the urine can also be measured. Exogenous antidiuretic hormone (ADH) is given to the patient, and the urine osmolarity is measured 1 hour later. The administered ADH should cause the urine to be more concentrated by increasing renal water retention. If retention is impaired, ADH does not increase the urine osmolarity.

Imaging techniques are also used to obtain information about the status of the urinary system. The system can be examined by CT and MRI scans, ultrasound, angiography, and techniques in which the urinary tract is infused with a radiopaque substance to make it visible by X ray. These X-ray techniques include the following:

- In **intravenous urography,** an iodinated benzoic acid derivative is administered intravenously. The kidney rapidly filters the chemical, and the iodine makes the image more opaque on an X ray, so that the kidney and lower urinary tract can be viewed clearly. This technique is often used to locate sites of renal injury.

- **Cystography** uses a radiopaque agent infused by catheter into the bladder. This technique allows for controlled filling of the bladder with the agent and consequently a clearer image than is possible with intravenous urography. It is most often used to diagnose neurogenic bladder, a ruptured bladder, or cystitis.

- **Retrograde pyelography** employs a radiopaque chemical infused through a catheter until it fills the renal calices, renal pelvis, ureters, and bladder. Because this technique does not require the radiopaque agent to be filtered by the kidney, it provides a more detailed image, especially of the lower urinary tract (urinary bladder and urethra).

In addition to imaging techniques, renal biopsy can be used to confirm a diagnosis.

Urinary Tract Infections: Cystitis and Pyelonephritis

Urinary tract infections (UTIs) may occur in any portion of the urinary tract. Bacteria (*Escherichia coli*, *Pseudomonas*, *Klebsiella*, *Staphylococcus*, and *Proteus*) are the most commonly encountered microbial agents. These bacteria usually move from the perineum to the urethral orifice and then progress in a retrograde manner into the urethra, urinary bladder, ureters, and finally the kidneys. People most at risk of UTI include prepubertal children, the elderly, sexually active women, pregnant women, people who practice poor hygiene, and individuals diagnosed with neurogenic bladder, urinary tract obstruction, or diabetes mellitus. It has been estimated that at least 10% to 20% of all women in the United States have lower urinary tract infections at any one time.

Several physiological factors limit the occurrence of urinary tract infections, including the bactericidal effects of urea, the acidic pH of the urine, the "washing out" of bacteria during micturition, and closure of the urethral openings of the bladder during micturition, thus minimizing urine reflux. In men, the length of the urethra and secretions from the prostate also reduce the potential for UTI.

The most common UTI is **cystitis**—bladder inflammation accompanied by dysuria, increased urinary frequency and urgency, and pain in the pubic and lumbar regions. The urine may also appear cloudy due to the presence of bacteria (**bacteriuria**) or pus (**pyuria**). If infection causes bleeding in the bladder, **hematuria** may also occur. Severe or prolonged infections may induce ulceration of the bladder *(ulcerative cystitis)* or death of bladder cells *(gangrenous cystitis).* Cystitis is diagnosed through the patient history and urinalysis. The most common treatment is antibiotics, but in cases where urinary tract obstruction is the suspected cause, the obstruction must be treated as well.

Pyelonephritis, infection of the renal pelvis and medulla, is another form of UTI. Most often, the same bacteria that cause cystitis cause pyelonephritis; however, this disease may also be caused by a virus or fungus. Pathogens may invade the kidneys when urinary obstruction causes a backflow of urine from the bladder to the kidneys, or they may invade by way of the blood. One or both kidneys may be involved. As with cystitis, most cases of pyelonephritis occur in women. As the infection intensifies, abscesses and necrosis may occur within the kidney, evidenced by pyuria and leading to scarring of the kidney and atrophy of renal tubules.

Signs and symptoms of pyelonephritis include fever, pain in the groin and/or flank, dysuria, and increased urinary frequency. Diagnosis of pyelonephritis is through urinalysis and urine culture, since it is often difficult to distinguish between cystitis and pyelonephritis based on clinical symptoms alone. Urinalysis reveals leukocyte-containing casts and more leukocytes than would be present in cystitis. Urine culture shows more antibody-coated bacteria than would be identified in cystitis.

Pyelonephritis is treated the same as cystitis, except that antibiotics specific for the causative agent are used and the duration of antibiotic treatment is longer. In addition, the patient's urine is rechecked through cultures at 1 and 4 weeks after treatment, especially if symptoms recur. This precaution is due to the increase in the numbers of antibiotic-resistant bacteria and the possibility of more severe damage to the kidneys.

Glomerulopathies: Glomerulonephritis and Nephrotic Syndrome

Glomerulopathies are diseases that affect mainly the function of the glomerulus. These disorders may be the primary disease, or they may result secondarily from some other systemic disease. All cases are distinguished by damage to the glomeruli with changes in capillary permeability. The two most common glomerulopathies are *glomerulonephritis* and *nephrotic syndrome,* both of which can affect people of any age, sex, or ethnic group.

Glomerulonephritis is an inflammation of the glomeruli due to infection, systemic disease, drugs, toxins, or an immune disorder. Due to the wide variety of causes, glomerulonephritis is often characterized by the specific causative agent or the progression of the disease (acute, chronic, or rapidly progressive). In general, the signs and symptoms of all types are similar, differing only slightly. A general discussion of glomerulonephritis is presented here.

Patients with glomerulonephritis usually show marked changes in urine composition due to changes in glomerular permeability. They typically exhibit proteinuria, hematuria, leukocyturia, and casts. Other signs include edema, systemic hypertension, and elevated *blood urea nitrogen (BUN).* Renal function tests reveal a decreased glomerular filtration rate. Renal biopsy is often performed to determine the type and extent of injury to the kidneys.

Treatment of glomerulonephritis centers on the causative agent and managing the accompanying dysfunction. Although each type of glomerulonephritis would have a specific treatment regimen, all are aimed at controlling edema, hypertension, hyperkalemia, and hyperlipidemia. Acute glomerulonephritis leads to death in 2% to 5% of patients, while patients with chronic glomerulonephritis may live productive lives for up to 20 years after diagnosis.

Nephrotic syndrome is a kidney disease defined by the excretion of more than 3.5 g of protein per day in the urine. It is often accompanied by an abnormally low level of serum albumin. Since plasma protein level is a major factor in the osmotic uptake of water from the tissue fluid, this state of hypoproteinemia typically results in edema and ascites. Thus, a triad of proteinuria, hypoproteinemia, and edema is the chief sign of nephrotic syndrome. Nephrotic syndrome has a variety of forms and causes. It can be congenital, or it can be triggered by glomerulonephritis, certain drugs (such as street heroin), certain infections (such as HIV and some tropical parasitic infections), and some systemic diseases (such as diabetes mellitus and systemic lupus erythematosus).

Patients may first notice foamy or frothy urine. Subsequently, the patient history and physical examination often reveal anorexia, edema, abdominal pain, and muscle wasting. The edema may be localized to a single body region and may lead to other signs and symptoms such as chest pain and dyspnea, stemming from pleural effusion. Edema often appears in the face in the morning and in the knees or ankles later in the day. Nephrotic syndrome is suspected when the aforementioned triad of signs appears, and is confirmed by additional blood and urine findings, including elevated levels of serum lipids, fatty casts and blood cells in the urine, and urinary excretion of more than 3.5 g of protein per day. A final confirmation often comes from a renal biopsy showing damage to the glomerular basement membrane and the presence of autoantibodies in the glomerulus. Nephrotic syndrome can lead to renal failure and death. Some other consequences of nephrotic syndrome include susceptibility to infection due to the loss of immunoglobulins from the blood, prolonged blood clotting due to the loss of clotting factors, iron-deficiency anemia due to the loss of iron-transport proteins from the blood, and atherosclerosis due to the elevated serum lipid levels.

Treatment of nephrotic syndrome is twofold, involving first treating the causative agent and then the associated symptoms. The most common treatments include dietary restrictions (low salt and low fat), protein supplements to restore the plasma albumin level, anti-inflammatory drugs (most often steroids), and diuretics to control edema or hypertension. If damage to the kidney is severe, lifelong hemodialysis may be required.

Neurogenic Bladder

Neurogenic bladder results when the innervation of the bladder is disrupted. The disruption can be at the level of the brain, spinal cord, or peripheral nerves supplying the bladder. Disruption of the micturition reflex leads to incontinence, increased urination frequency, or a urinary obstruction that causes retention of urine within the bladder. There are a wide variety of causes of neurogenic bladder, including congenital abnormalities, trauma such as a ruptured intervertebral disc, and diseases such as syphilis, diabetes mellitus, central nervous system tumors, demyelinating or degenerative nervous system disease, or a cerebrovascular accident. However, the most common cause is spinal cord injury that results in paraplegia.

Although nervous control of the bladder is impaired in all cases of neurogenic bladder, the effect of the disruption is not always the same. For example, in the absence of nervous regulation, the bladder may be in either a hypotonic (flaccid) or spastic (contracted) state. In general, congenital defects produce hypotonic bladder, while trauma and disease produce either hypotonic or spastic bladder.

The signs and symptoms of neurogenic bladder vary due to the wide variety of causes. Diagnosis involves imaging techniques such as intravenous urography, ultrasound, and cystography, as well as assessment of the micturition flow rate. Treatment depends on the cause. Many patients are treated with catheterization and medications to improve bladder control. Patients must be continually monitored for renal function, infection, and the formation of calculi. Limiting dietary calcium intake minimizes calculus formation. If not treated and managed properly, neurogenic bladder can cause other renal disorders and impair renal function.

Using Diuretics to Treat Urinary System Disorders

Diuretics are chemicals that increase urine volume. They are usually used to reduce total body water and blood pressure. Thus, they are helpful in controlling hypertension, edema, congestive heart failure, and some other diseases. Diuretics are divided into different classes based on their mechanism of action:

- **Aldosterone antagonists** These are agents that work on the distal convoluted tubule to block the sodium-retaining action of aldosterone. Thus, more sodium remains in the renal tubule and passes in the urine, and where sodium goes, water follows.

- **Carbonic anhydrase (CAH) inhibitors** CAH is employed in the proximal convoluted tubules to break carbonic acid down into H^+ and $HCO3^-$. The H^+ is then normally excreted in exchange for Na^+. By blocking the action of CAH, CAH *inhibitors* reduce H^+ secretion and Na^+ retention. Thus, again, more sodium and more water pass in the urine.

- **Osmotic diuretics** These are drugs that are freely filtered by the glomerulus but not reabsorbed by the renal tubules. Thus, they remain in the tubules, increasing the osmolarity of the tubular fluid. This osmolarity opposes tubular reabsorption of water, so more water is passed in the urine.

- **Sodium and chloride reabsorption inhibitors** These diuretics work mainly in the nephron loop and are therefore also called *loop diuretics.* They block the reabsorption of Na^+, K^+, and Cl^- and thus promote salt and water elimination.

Each of these diuretics is appropriate to the treatment of different diseases, and they vary in their side effects. Some may cause hyperkalemia and others hypokalemia; some cause alkalosis and others acidosis; and so on. Before prescribing any of them, the clinician must have a thorough understanding of the disease at issue, the patient's overall physiological condition, and the mode of action and side effects of each diuretic.

Case Study 23 The International Student with Renal Disease

Haddi is a young Nigerian woman studying in the United States. One afternoon in March, she reports to the university health center complaining that she doesn't feel well, she has no appetite, and her stomach hurts. She is told to come back first thing the next morning to give a urine specimen and have a physical examination. That day, her urine specimen is oddly frothy, and the nurse notes that Haddi's eyelids are puffy. A dipstick urinalysis shows a high concentration of protein in the urine. Since this indicates a possibly serious disorder, the clinic refers Haddi to a urologist.

In taking Haddi's history, the urologist learns that Haddi often notices that her face is puffy in the morning, and by afternoon she frequently has swelling in the knees and ankles. She occasionally has abdominal pains and sometimes difficulty breathing. The urologist asks Haddi about her travel history and history of other illnesses. Haddi says that she goes home to Nigeria during Christmas and summer breaks, and that she almost always gets malaria when she is there. She last went home in December, and had a bout of malaria then, as usual, but obtained treatment and her symptoms (chills and fever) disappeared. The urologist admits Haddi to the hospital for overnight observation and a 24-hour urine collection. Some of the results of her physical examination and laboratory work are shown here.

Vital signs:

Oral temperature = 98.6°F (37.0°C)
Heart rate = 68 beats/min
Respiratory rate = 24 breaths/min
Blood pressure = 131/73 mmHg

Physical examination:

Edema of lower limbs, mild ascites

Blood:

Hematocrit (Hct) = 34%
RBC count = 3.3 x $10^6/\mu L$
Total protein = 3.1 g/dL
Albumin = 1.6 g/dL
Sodium = 136 mEq/L
Other serum electrolytes = Normal
Blood urea nitrogen (BUN) = 57 mg/dL (mild azotemia)
Lipids: Fat droplets present
Low-density lipoproteins = 220 mg/dL
Triglycerides = 165 mg/dL
Cholesterol = 238 mg/dL

Urine:

pH = 5.5
Specific gravity = 1.052
Protein excretion = 15.5 g/day
Glucose and ketones = Negative
Appearance: Light yellow, frothy.
Urine culture: No pathogenic microorganisms.
Sediment shows fatty casts, RBCs, and WBCs.
Dipstick tests show proteinuria and hematuria.

On the basis of these findings and with Haddi's consent, the urologist orders a renal biopsy. The histopathologist observes disruption of the glomerular basement membranes, and a stain for immunoglobulins in the glomerulus is positive.

The urologist diagnoses Haddi with nephrotic syndrome. He explains to her that nephrotic syndrome can be triggered by certain forms of malaria, and often develops a few months after a malarial attack. He says that her blood work shows no signs of malarial parasites at present, and Haddi says she has not had any of the fever and chills of malaria since returning to school for the semester. The physician advises her that nephrotic syndrome often clears up when the underlying cause is successfully treated, as her malaria appears to be. He warns her, however, that repeated bouts of malaria can worsen the condition and cause potentially fatal renal failure, and furthermore that malaria sometimes does not yield to drug therapy in people with nephrotic syndrome. These facts make it critically important, he says, that she take extreme measures to avoid malaria-carrying mosquitoes when she goes home and that she carefully observe malaria prophylaxis—taking drugs in advance of her trips home to prevent malaria infection even if she is bitten.

In the meantime, the physician advises that Haddi remain in the hospital for treatment. She receives furosemide, a diuretic to treat her edema; an immunosuppressant to control the immune attack on her glomeruli; and I.V. albumin. She is placed on a low-fat, low-salt diet. From March through May, Haddi's serum albumin returns to a normal level of 3.5 g/dL, her urinary protein excretion declines to a low level, and she is gradually withdrawn from the diuretic and immunosuppressant. Before traveling home in May, she takes a regimen of chloroquine for protection against malaria.

Based on this case study and other information in this chapter, answer the following questions.

1. Nephrotic syndrome is sometimes caused by diabetes mellitus. How do we know this is not the cause in Haddi's case?
2. In nephrotic syndrome, what accounts for the froth in a freshly collected urine specimen?
3. Which data obtained from Haddi's blood and urine are especially consistent with the edema she experiences?
4. Explain the pathophysiological reasons that Haddi has ascites, azotemia, and hematuria.
5. Why is Haddi given intravenous albumin? Which of her symptoms would be relieved by this treatment?
6. Aside from malaria prophylaxis, what are some other protective measures Haddi could take on her trips home in order to reduce her risk of kidney failure?
7. Howard, a 65-year-old male, is prescribed an osmotic diuretic, mannitol, for the treatment of hypertension. Explain how mannitol would affect his blood pressure and his daily urine output.-
8. Based on your knowledge of the role of the renal tubule in regulating ion balance, explain how a diuretic could induce hyperkalemia. Then explain how a different diuretic might induce hypokalemia.
9. Susan, a 12-year-old girl, is brought to her pediatrician for a routine physical. Her mother mentions that Susan seems to be drinking a lot more water than normal. Urinalysis reveals elevated specific gravity, decreased pH, and glycosuria. Which of the following clinical signs would support a diagnosis of diabetes mellitus?
   a. ketonuria
   b. pyuria
   c. oliguria
   d. hemoglobinuria
   e. bright yellow urine
10. For each of the four diseases listed below, you are given a choice of two diagnostic methods. State which of the two methods you would use and why.

    cystitis: retrograde pyelography or urine specific gravity?

    pyelonephritis: urine culture or renal clearance test?

    diabetes mellitus: chemical urinalysis or renal biopsy?

    nephrotic syndrome: urine culture or chemical urinalysis?

Selected Clinical Terms

anuria A state of severely reduced urine output, under 100 mL/day, presenting a threat of azotemia and uremia.

cast A cylindrical, solid mass found in the urine sediments, molded by the renal tubule and composed of protein and various combinations of blood cells, epithelial cells, fat, bacteria, and crystals; typically an indication of serious renal pathology.

dysuria Pain upon urination.

enuresis Incontinence; the inability to retain urine or control its elimination.

hematuria Blood in the urine.

nocturia Having to urinate during the night.

oliguria A state of inadequate urine output, under 500 mL/day.

polyuria A state of excessive urine output, exceeding 2 L/day.

pyuria Pus in the urine.

24 Water, Electrolyte, and Acid-Base Balance

Objectives

In this chapter we will study

- various forms of diabetes, especially gestational diabetes and diabetes insipidus;
- abnormalities in sodium balance—hypernatremia and hyponatremia;
- abnormalities in potassium balance—hyperkalemia and hypokalemia; and
- the diagnosis of various forms of acid-base imbalance.

Diabetes

When people say "diabetes," they usually mean diabetes mellitus. There are other, less common forms of diabetes, however, that have nothing to do with insulin. The one thing they have in common is an abnormally high volume of urine output; the word **diabetes** means "passing through," and diabetics of all kinds pass a great deal of water. A general definition of diabetes, embracing all its varieties, is chronic polyuria resulting from a metabolic disorder. (Thus, it does not include the temporary polyuria induced by drinking a lot of water or beer.) In this chapter, we examine the other forms of diabetes in a little more depth.

Diabetes Mellitus

Diabetes mellitus (DM) results from the hyposecretion of insulin (type I, or IDDM) or target cell insensitivity to insulin (type II, or NIDDM). Its signs include polyuria, polydipsia (intense thirst), polyphagia (intense hunger), hyperglycemia (elevated blood glucose), glycosuria (glucose in the urine), and ketonuria (ketones in the urine). These conditions have numerous devastating effects on the body, leading to such consequences as blindness, gangrene, renal failure, and early death if untreated.

Gestational Diabetes

Gestational diabetes (GDM) is a form of diabetes mellitus that develops in pregnant women. It is also called type III diabetes mellitus. Overall, it occurs in 1% to 3% of pregnant women, but the incidence is significantly higher in some groups, including people of Mexican, Indian, American Indian, Asian, and Pacific Islander descent. Pregnancy reduces a mother's insulin sensitivity as a way of "sparing" blood glucose for the nourishment of the fetus. Gestational diabetes occurs when this mechanism overcompensates for the needs of the fetus, so the mother experiences hyperglycemia and glycosuria, and of course the glycosuria osmotically produces polyuria. Unrecognized and untreated GDM is a common cause of fetal deformity and miscarriage, so it is advisable to screen all pregnant women for GDM. When diagnosed, GDM is managed with dietary modification, weight control, exercise, and small doses of insulin. Also, fetal development is not allowed to go beyond full term (42 weeks) because of the high rate of mortality in overdue fetuses. Newborns with diabetic mothers are at high risk of numerous disorders and are therefore given extra forms of neonatal assessment. Even when a woman has preexisting DM, it places the life of her fetus in jeopardy, and the treatment of DM in pregnancy is a very delicate matter.

Diabetes Insipidus

Diabetes insipidus results from the hyposecretion or inaction of antidiuretic hormone (ADH). Physicians used to taste a patient's urine to test for glucose; thus the name diabetes insipidus refers to the urine's lack of sweetness. **Central diabetes insipidus** results from a lack or deficiency of ADH, usually as a result of cranial trauma or surgical removal of the pituitary. **Nephrogenic diabetes insipidus (NDI)** results from a lack of ADH receptors in the collecting ducts of the kidney, thus making ADH ineffective. NDI is sometimes hereditary but more often a side effect of certain antibiotics, anesthetics, and other drugs. The hereditary form affects principally males and is thought to be an X-linked recessive trait. This form develops shortly after birth. Its signs include abnormally frequent urination and increased feeding frequency, the latter owing to the baby's dehydration and intense thirst. Since the urine is hypotonic, the body's sodium concentration becomes elevated. If the parents do not recognize the condition, this hypernatremia can lead to vomiting, fever, convulsions, and sometimes brain damage and permanent mental retardation. Even once the disease has been diagnosed, a child grows slowly because of episodes of dehydration throughout infancy and childhood.

Both forms of diabetes insipidus are diagnosed with a water deprivation test. Water is withheld from the patient for 12 to 14 hours. Then the patient is allowed to drink, and the osmolarity of the subsequent urine samples is measured. Patients with diabetes insipidus have an extremely low urine osmolarity (less than 200 mOsm/L) and specific gravity (1.00 to 1.005). NDI is further diagnosed from the fact that this osmolarity shows little or no change even when the patient is given ADH. The highest priority in treatment is to replace the lost water and restore the patient's fluid volume and osmolarity. Diabetes insipidus is treated essentially the same as hypernatremia, described next. A child with hereditary NDI must be taught about the condition at an appropriate age and cautioned to drink ample water.

Abnormalities in Sodium Balance

This section and the next provide insights into the pathology of sodium and potassium excesses and deficiencies. Because of the especially critical role of sodium and potassium in the heartbeat, other muscle contractions, and nerve function, these two ions command the greatest attention in discussions of electrolyte balance.

Hypernatremia

Hypernatremia is a serum Na^+ concentration above 145 mEq/L. One cause of hypernatremia is the increased intake or retention of sodium. The source of excess sodium is rarely dietary, but more often fluid therapy—for example, I.V. sodium bicarbonate. Hypernatremia can also result from excess sodium retention owing to the hypersecretion of aldosterone. This is sometimes a secondary effect of ACTH hypersecretion, as in Cushing syndrome. The other basic cause of hypernatremia is insufficient water intake or excessive water loss without a proportionate loss of sodium. Excessive water loss can have several root causes: sweating, diabetes insipidus, the use of loop and osmotic diuretics, infant diarrhea, respiratory loss (especially when a respiratory infection or fever induces *tachypnea,* or accelerated breathing), a blunted sense of thirst, lack of access to potable water (as in exposure at sea or in a desert), infirmity or immobility (inability to get a drink for oneself), and coma (resulting in the inability to move or express the need for water). Hypernatremia is especially common in elderly hospital patients, who suffer high mortality as a result. A caregiver must be especially attentive to the needs of patients who cannot easily move or express their needs.

The signs and symptoms of hypernatremia result mainly from the hypertonic state of the extracellular fluid and resulting shrinkage of nerve cells. They include confusion, neuromuscular excitability, seizures, and coma. Hypernatremia is diagnosed by the measurement of serum Na^+ concentration, a high urine specific gravity (> 1.030), elevated hematocrit, and elevated concentration of plasma protein. It is treated primarily by giving water or 5% dextrose in water, orally if a patient is conscious and able to swallow and intravenously otherwise. This dilutes the sodium and restores normal osmolarity. However, both oral and intravenous rehydration must be done gradually (over 48 hours) to prevent excessively rapid rehydration and cerebral edema. Diuretics are sometimes given to promote Na^+ excretion, often with isotonic dextrose and KCl to compensate for the fluid loss.

Hyponatremia

Hyponatremia is a serum Na^+ concentration below 135 mEq/L. One cause is reduced renal excretion of water, as seen in renal failure, congestive heart failure, and cirrhosis. The kidneys retain more water, which dilutes the Na^+ of the extracellular fluid. Another cause is excessive Na^+ loss—for example, through vomiting, diarrhea, sweating, and burns—and replacement of the lost fluid with plain water (producing **dilutional hyponatremia**).

The signs and symptoms of hyponatremia stem largely from the swelling of neurons as they take up fluid from the hypotonic ECF. Neurological signs include lethargy, headache, confusion, stupor, neuromuscular excitability, seizures, and coma. Nonneurological signs include weight gain, edema, ascites, and distended jugular veins. Note that the neurological signs are similar to those for hypernatremia, so they alone are insufficient for a final diagnosis. Differential diagnosis is achieved by a finding of low serum Na^+ concentration and a low urine specific gravity (< 1.010). Hyponatremia is treated by restricting water intake, replacing sodium, and correcting the underlying cause. Hyponatremia sometimes results from drugs such as thiazide diuretics, and may require a change of medication.

Abnormalities in Potassium Ion Balance

Whereas sodium is the primary extracellular cation, potassium is the primary intracellular cation. As little as 2% of the body's potassium is found in the extracellular fluid; most is found in skeletal muscle. In fact, total body potassium can be used as an estimator of lean body mass. Even though most potassium is within the cells, changes in serum potassium adequately measure the body's potassium balance unless the individual has a disease that alters membrane function, total body mass, or the acid-base balance of the body.

Hyperkalemia

Hyperkalemia is a serum potassium concentration above 5.0 mEq/L. This condition is relatively rare because the kidneys are very efficient at excreting excess K^+, but it can result from excessive potassium intake (for example, in patients on I.V. fluid therapy or people who overuse potassium salt substitutes), inadequate potassium excretion (in renal failure or aldosterone hyposecretion), or maldistribution of potassium between the intracellular and extracellular fluids (ECF and ICF). Maldistribution can come about by several means. Normally, 98% of the body's potassium is in the ICF. Hemolytic anemia, massive crush injuries, burns, extensive surgery, and transfusion with stored blood (containing old, leaky RBCs) can release large quantities of K^+ from the ICF into the ECF and cause sudden, potentially fatal hyperkalemia. Acidosis typically induces hyperkalemia because excess H^+ enters cells and K^+ exits them to compensate, thus raising the K^+ concentration in the ECF. Insulin promotes K^+ uptake by cells, so an insulin deficiency can also cause excess K^+ to accumulate in the ECF. Medications such as potassium-sparing diuretics, β-blockers, NSAIDS, and ACE inhibitors also sometimes cause hyperkalemia.

Hyperkalemia can produce confusing and seemingly contradictory signs and symptoms, depending in part on how rapidly the K^+ concentration rises. Common effects include muscular weakness, neuromuscular excitability, and in severe cases, ventricular fibrillation and cardiac arrest. A rather grim application of this fact is that high-potassium injections are used in veterinary euthanasia and in capital punishment by lethal injection.

Hyperkalemia can be diagnosed from a combination of patient history (for example, trauma, blood transfusion, insulin deficiency, or Addison disease), characteristic abnormalities of the electrocardiogram, and measurement of serum K^+ concentration. Mild hyperkalemia can be treated with dietary modification and, if the disorder is a side effect of medication, by changing medications. Insulin and glucose can be given to lower the K^+ level by inducing cellular uptake; drugs are available to reduce neuromuscular excitability until the K^+ concentration is restored to normal; and hemodialysis is used if the hyperkalemia results from renal failure.

Hypokalemia

Hypokalemia is a serum K^+ concentration of less than 3.5 mEq/L. The fundamental causes of hypokalemia are inadequate intake, excessive loss, or maldistribution of potassium between the ECF and the ICF. Dietary deficiency is rare, but hypokalemia is not uncommon in people with depressed appetites and poor nutrition, as occurs in alcoholism and anorexia nervosa. Excessive losses of K^+ can be due to heavy sweating, vomiting, diarrhea, and laxative abuse. Diarrhea can increase the fecal loss of K^+ from the normal rate of 5 to 10 mEq/day to as much as 200 mEq/day. Excessive urinary loss can result from aldosterone hypersecretion. (Aldosterone promotes Na^+ retention and K^+ excretion.) Hypokalemia also occurs if an excess of K^+ transfers from the ECF to the ICF. In alkalosis, for example, H^+ ions diffuse out of the cells into the ECF, and K^+ diffuses from the ECF into the cells to replace the H^+. Thus, the ECF concentration of K^+ falls below normal. Insulin can induce severe and even fatal hypokalemia if a patient takes it without also taking potassium supplements. The reason is that diabetic ketoacidosis causes H^+ to enter cells, and K^+ leaves the cells to compensate for the H^+ inflow. The K^+ that leaves is excreted in the urine, so the ECF K^+ concentration remains normal, but the body's total K^+ stores are depleted. Then, when insulin is given and the ketoacidosis is corrected, K^+ re-enters the cells and the ECF becomes hypokalemic.

Hypokalemia makes cells less excitable. This is reflected in such clinical signs as muscle weakness, loss of muscle tone, depressed reflexes, and irregular heartbeat. Involvement of the respiratory muscles can bring about depressed ventilation or even respiratory arrest. A loss of smooth muscle tone produces constipation, nausea, vomiting, and intestinal bloating.

Hypokalemia can be diagnosed from the patient's history, characteristic alterations of the electrocardiogram, and measurement of the serum K^+ concentration. It is treated by correcting the causes of potassium loss and having the patient eat potassium-rich foods such as bananas and vegetables. If necessary, I.V. potassium can be administered, but it must be given slowly because potassium irritates blood vessels and because overly rapid administration of I.V. potassium can induce a dangerous state of hyperkalemia.

Diagnosing Acid-Base Imbalances

Acidosis and alkalosis can be either respiratory or metabolic in origin. For example, emphysema causes *respiratory acidosis* because the diseased lungs cannot expel CO_2 as fast as the body produces it. Diabetes mellitus produces *metabolic acidosis* because incomplete fat oxidation generates ketones (keto acids). Hyperventilation causes *respiratory alkalosis* because the lungs expel CO_2 faster than the body produces it. Chronic vomiting causes *metabolic alkalosis* because stomach acid is lost from the body.

When there is a respiratory dysfunction that upsets acid-base balance, the kidneys can sometimes *compensate* for it and restore homeostasis; if the kidneys do not respond, the condition is *uncompensated.* Conversely, if the acid-base imbalance is caused by a renal or metabolic disorder, the respiratory system may or may not compensate for it (by adjusting respiratory rate); thus we can have **compensated** or **uncompensated** metabolic acidosis or alkalosis. This very simplified summary does not consider conditions beyond the scope of this manual, such as *mixed disturbances* in which a person might have both metabolic alkalosis and respiratory acidosis. But this introduction at least gives you a general idea of the reasoning process used in diagnosing acid-base disorders.

An important aspect of the clinical assessment of an acid-base imbalance is to determine its origin (respiratory or metabolic) and whether it is compensated or uncompensated. This informs a clinician of what underlying condition may call for treatment, and whether the body is restoring homeostasis on its own or whether clinical intervention (such as I.V. fluid) is necessary to restore a normal pH. Such determinations can be made from measurements of the pH of arterial blood and the PCO_2 and HCO_3^- concentration of venous blood.

Table 24.1 shows the normal values of these three variables and what is to be expected in each major form of acid-base imbalance. It is important to see beyond the table entries, however, to the physiological rationale for each, as explained after the table. If you understand the rationale, you can fill in such a table by reason rather than rote memorization.

Table 24.1 Forms of Acid-Base Imbalance

Normal Values	pH 7.35–7.45	PCO_2 35–45 mmHg	HCO_3^- 22–26 mEq/L
Acidosis			
Compensated respiratory	Reduced	Elevated	Elevated
Uncompensated respiratory	Reduced	Elevated	Normal
Compensated metabolic	Reduced	Reduced	Reduced
Uncompensated metabolic	Reduced	Normal	Reduced
Alkalosis			
Compensated respiratory	Elevated	Reduced	Reduced
Uncompensated respiratory	Elevated	Reduced	Normal
Compensated metabolic	Elevated	Elevated	Elevated
Uncompensated metabolic	Elevated	Normal	Elevated

Table 24.1 shows eight major possibilities:

1. **Compensated respiratory acidosis** The pH is low by virtue of the definition of acidosis. The PCO_2 is high because the lungs are not expelling CO_2 as fast as the body produces it. CO_2 acidifies the blood, so this is the cause of the acidosis. The kidneys are attempting to compensate for this acid load by retaining bicarbonate (which buffers acid), so the bicarbonate level is elevated. This situation is typical of long-term pulmonary dysfunctions such as emphysema and pneumoconiosis (see chapter 22 of this manual).
2. **Uncompensated respiratory acidosis** The pH is low and the PCO_2 is high for the reasons stated in *1*. The bicarbonate level is *not* elevated, however, indicating that the kidneys are not compensating for the respiratory dysfunction. This condition may be seen in cases of short-term asphyxia or holding the breath, letting CO_2 accumulate but not allowing enough time for renal compensation to take effect.
3. **Compensated metabolic acidosis** The pH is low because of some metabolic acid load. The PCO_2 is low because the respiratory system is "blowing off" CO_2 faster than the body produces it, attempting to compensate for the acidosis. This is typical of conditions like diabetes mellitus, which loads the body with acidic ketone bodies.
4. **Uncompensated metabolic acidosis** The pH and bicarbonate concentrations are low, but there is nothing unusual in the PCO_2 because the respiratory system is not compensating for the imbalance. This could occur in response to the use of an acidic drug or to fluid and electrolyte imbalances that put the body into a state of acidosis, accompanied by an inability of the respiratory system to adjust the ventilation rate (for example, because of depression of the respiratory center by drugs or anesthesia, or pulmonary diseases that reduce lung function).
5. **Compensated respiratory alkalosis** The pH is high (by definition of alkalosis), and the PCO_2 is low because the respiratory system is expelling CO_2 faster than the body produces it. To compensate, the urinary system excretes extra bicarbonate, so the blood bicarbonate concentration is reduced.
6. **Uncompensated respiratory alkalosis** The pH is high and the PCO_2 is low, but the bicarbonate level is normal because the urinary system is not compensating for the respiratory dysfunction. This is typical of hyperventilation, which rapidly expels CO_2 but does not last long enough to activate renal compensation.
7. **Compensated metabolic alkalosis** The pH is high by definition, and the PCO_2 is high because the respiratory system is retaining CO_2 (not expelling it as fast as it is produced). CO_2 lowers the blood pH and compensates for the metabolic condition. This may occur in chronic vomiting, as in pregnant women with severe morning sickness *(hyperemesis gravidarum)*.
8. **Uncompensated metabolic alkalosis** The pH is high, but the PCO_2 is normal; the respiratory system has not adjusted CO_2 elimination to compensate for the metabolic dysfunction. This could also result from chronic vomiting, among other disorders, accompanied by such compromises in pulmonary function as noted in *4*.

Case Study 24 The Very Thirsty Baby

Charles, a 4-month-old Asian boy, is brought to his pediatrician by his mother. Charles has two older sisters, and his mother has noticed that he is nursing much more frequently than they did. She thinks this heavy nursing is the cause of the excessive amounts of urine he is producing. She reports that he wants to nurse every hour, and she has to change his diaper about every 30 minutes. In addition, because of the frequency of nursing, she is using an infant formula to supplement her milk in order to meet Charles's demands.

When reviewing the records of Charles's two sisters, the pediatrician notes that none of these symptoms were reported for either sister. The pediatrician completes a physical examination and notes the following:

Vital signs:

Rectal temperature = 98.9°F (37.2°C)

Heart rate = 85 beats/min

Respiratory rate = 13 breaths/min

Has lost some weight since last visit.

Reflexes: All within normal range, but slightly excitable.

Skin and mucous membranes: Lack of skin turgor; sunken fontanels; mucous membranes dry.

Based on the results of the physical examination, the pediatrician asks to take both blood and urine samples for analysis. The mother agrees, and the results are as follows:

Blood:

Hematocrit (Hct) = 59%

Serum sodium = 139 mEq/L

Serum potassium = 5 mEq/L

Serum bicarbonate = 21 mEq/L

Serum chloride = 109 mEq/L

Urine:

Specific gravity = 1.001

pH = 6.8

Glucose, protein, lipids, blood all absent

Based on these results, the pediatrician suspects that Charles's kidneys are not correctly conserving water. He orders another blood test, which shows an elevated ADH level, and renal function tests, which show a normal glomerular filtration rate and renal plasma flow. Suspecting that Charles is suffering from nephrogenic diabetes insipidus, the pediatrician orders a water deprivation test, which shows no change in urine osmolarity over the course of the test. This indicates lack of ADH responsiveness by the kidneys. Based on these results, Charles is diagnosed with nephrogenic diabetes insipidus.

Based on this case study and other information in this chapter, answer the following questions.

1. Why is nephrogenic diabetes insipidus not seen in Charles's sisters? Why is it not seen in either of his parents? If Charles later has a baby brother, would you expect the baby to have this disorder? Explain.
2. What aspects of the history, physical examination, and laboratory tests support the diagnosis of nephrogenic diabetes insipidus?
3. If you were the pediatrician and Charles's mother asked you if he was going to need insulin injections and dietary restrictions, what would you tell her?
4. As Charles gets older, what type of counseling should he receive to help him control this disorder?
5. What will be the consequences if Charles disregards this counseling?
6. Why do some patients with nephrogenic diabetes insipidus have *elevated* levels of ADH?
7. With respect to its pathogenesis, does nephrogenic diabetes insipidus more nearly resemble type I or type II diabetes mellitus? Explain.
8. A patient with hyponatremia would be given
   a. rapid infusion of hypotonic saline.
   b. rapid infusion of hypertonic potassium chloride.
   c. slow infusion of hypertonic saline.
   d. slow infusion of hypotonic potassium chloride.
   e. antidiuretics.
9. Jack goes into cardiac arrest but is revived in 3 minutes by the paramedics. They administer intravenous sodium bicarbonate to correct the acidosis that developed during the cardiac arrest. He then begins to exhibit excessive neuromuscular excitability and starts to have seizures. Explain (*a*) why his cardiac arrest produced a state of acidosis, (*b*) why sodium bicarbonate would correct the acidosis, (*c*) why he developed hypernatremia, and (*d*) what could be done next to correct the hypernatremia.
10. Below are blood values from three patients. Identify which of the eight categories of acid-base balance each patient has.
    Patient A: pH = 7.62, PCO_2 = 55 mmHg, HCO_3^- = 32 mEq/L
    Patient B: pH = 7.25, PCO_2 = 48 mmHg, HCO_3^- = 34 mEq/L
    Patient C: pH = 7.10, PCO_2 = 42 mmHg, HCO_3^- = 8 mEq/L

Selected Clinical Terms

central diabetes insipidus Chronic polyuria due to hyposecretion of antidiuretic hormone.

compensated acidosis and alkalosis Acid-base imbalances in which the respiratory system adjusts pulmonary ventilation to compensate for a metabolic dysfunction or the kidneys adjust acid-base excretion to compensate for a respiratory dysfunction.

diabetes Any chronic polyuria of metabolic origin; when this word is used without a qualifier, it usually refers to diabetes mellitus.

dilutional hyponatremia A deficiency of sodium in the ECF resulting from excessive sodium loss and replacement of lost body fluids by ingestion of a hypotonic drink such as plain water.

nephrogenic diabetes insipidus Chronic polyuria due to renal insensitivity to antidiuretic hormone.

uncompensated acidosis and alkalosis Acid-base imbalances in which the respiratory system is unable to compensate for a metabolic dysfunction or the kidneys are unable to compensate for respiratory dysfunction, so that acid-base homeostasis cannot be restored without clinical intervention.

25 The Digestive System

Objectives

In this chapter we will study

- some symptoms of digestive system disorders;
- diagnostic techniques specific for the digestive system;
- pancreatitis;
- the liver disorders hepatitis and cirrhosis;
- three common inflammatory bowel diseases—ulcerative colitis, Crohn disease, and diverticulitis;
- appendicitis; and
- colorectal cancer.

Assessment of the Digestive System

This chapter discusses the techniques applied in diagnosing digestive system disorders and then goes on to describe some structural and functional diseases involving the digestive system. Disorders related to nutrition and metabolism are discussed in chapter 26.

Common Symptoms of Digestive System Disorders

Patients with disorders of the digestive system most commonly complain of such symptoms as pain, nausea, **dyspepsia** (indigestion), **anorexia** (loss of appetite), and **dysphagia** (difficulty swallowing). These symptoms can indicate disorders of other body systems as well, so it is necessary to take a careful patient history to determine whether the problem is digestive in nature. Helpful questions include, "When did you first notice the pain?" "Where do you feel the pain?" and "What makes the pain better (or worse)?" For example, a patient complaining of dyspepsia localized to the epigastric region and alleviated by eating may have an ulcer, while a patient experiencing abdominal pain in the right lower quadrant that is not alleviated by eating may have appendicitis.

Physical Examination

Examination of the digestive system involves observation, palpation, percussion, and auscultation. For example, if a patient complains of anorexia and dysphagia, the clinician may look into the oral cavity for lesions or bleeding gums. If the clinician notes **jaundice** (yellowish skin, corneas, mucous membranes, and body fluids), he or she may have uncovered a clue to liver disease or gallbladder obstruction. A distended abdomen could indicate anything from obesity or pregnancy to ascites or cancer. Striae (stretch marks) may suggest edema, Cushing disease, or ascites.

Palpating the abdomen enables the clinician to evaluate the size, shape, and texture of an organ. Intestinal peristalsis can be felt in nonobese patients, and abnormal peristaltic waves can indicate certain diseases. In many liver diseases, the liver becomes larger and firmer than normal. Palpation also allows the clinician to determine whether the abdominal muscles are excessively contracted or relaxed. In some diseases, a patient experiencing abdominal pain may contract the abdominal muscles to "protect" the region; this is known as **abdominal guarding**. In addition, some patients present with **rebound tenderness** whereby they do not experience excessive pain upon palpation but report sudden pain when the examiner's fingers are removed. Using palpation, the clinician may also be able to pinpoint the specific areas that are painful to help determine precisely which organ is involved as well as to check for abdominal masses, herniation of the abdominal wall, or distension of the abdomen.

Although percussion is routinely used to evaluate the respiratory system, it is less helpful in examining the gastrointestinal (GI) tract. There is normally little air in the GI tract to carry sound waves. However, percussion is useful in abdominal examinations to detect abnormalities of the liver and spleen and fluid and gas accumulation in the abdominal cavity or lower GI tract.

Auscultation is very useful in determining the peristaltic actions of the GI tract. Contractions of the smooth muscle of the tract produce characteristic

bowel sounds—normally 5 to 30 clicks or gurgles per minute. Prolonged gurgling (a "growling stomach") is called **borborygmus.** The frequency of the bowel sounds is reduced by intestinal obstruction and increased in diarrhea.

Laboratory Procedures

Following physical examination, various laboratory procedures can narrow down a diagnosis. For example, some liver disorders cause increased levels of hepatic enzymes in the blood serum (aspartate transaminase, alanine transaminase, alkaline phosphatase), reduce the concentrations of plasma protein, or increase the prothrombin time or clotting time. An increase in bilirubin may suggest liver or gallbladder disease.

Analysis of the feces (stool) also provides diagnostic information. The presence of blood suggests inflammation, ulceration, or cancer in the GI tract. When blood is present in quantities too small to be visible, it is called **occult blood.** Occult blood can nevertheless be detected by chemical tests. **Steatorrhea** (fatty feces) indicates pancreatic disease, cystic fibrosis, or a disorder affecting absorption. It is normal for the feces to contain certain bacteria, such as *E. coli.* If infection is suspected, however, the stool sample may be examined for parasites (such as amebae, tapeworms, or roundworms) or their eggs, or cultured for bacterial identification. *Salmonella, Campylobacter,* or *Shigella* is commonly found in cases of acute diarrhea. Finally, antibodies to hepatitis and other viruses and antigens diagnostic of colorectal cancer can also be detected in a stool sample.

Imaging Techniques

The digestive system can be examined by MRI, CT, and X ray (radiography). Radiography of the GI tract uses a contrast medium, such as barium, to make structures more visible. A *barium swallow* is used for visualization of the esophagus, stomach, and duodenum, and a *barium enema* for visualization of the colon. *Sialography* is radiography of the salivary glands; *cholecystography* is radiography of the gallbladder; and *cholangiography* is radiography of the bile ducts.

As mentioned in chapter 2 of this manual, endoscopy employs a flexible tube (endoscope) with a small fiber-optic camera to view interior spaces in the body. Its uses specific to the digestive system include *gastroscopy,* endoscopic examination of the stomach; *esophagogastroduodenoscopy,* examination of the tract from esophagus to duodenum; *colonoscopy,* examination of the large intestine; and *sigmoidoscopy,* examination of the rectum and sigmoid colon. In addition to providing a visual image, endoscopes can be equipped to obtain fluid or tissue samples for study.

Pancreatitis

Pancreatitis is inflammation of the pancreas. *Acute pancreatitis* is a severe inflammation requiring hospital care, but it is short-lived and leaves no lasting damage; *chronic pancreatitis,* by contrast, lasts longer and leaves the pancreas permanently damaged. Most people who develop pancreatitis are problem drinkers, but it can also be triggered by bile duct obstruction, infection, trauma, surgery, and other causes. The underlying mechanism is thought to be damage to the pancreatic acini and ducts, allowing potent pancreatic digestive enzymes to leak and digest the pancreatic tissue. As these enzymes invade the bloodstream, they damage the blood vessels and cause a loss of fluid into the tissues, thus presenting a threat of hypovolemia, circulatory shock, and renal failure.

A person with either form of pancreatitis experiences sudden epigastric and midabdominal pain, nausea, vomiting, sweating, fever, and tachycardia. Elevated levels of amylase and other serum enzymes, along with other hematologic signs, point to pancreatitis but are not specific enough to confirm this disease. A CT scan can aid diagnosis by revealing bile tract obstruction or pancreatic edema, and can also help reveal the extent of pancreatic damage. Pancreatitis that progresses to the point of necrosis and hemorrhage has a mortality rate of 10% to 50%. Treatment focuses on stopping the autodigestion of the pancreas, relieving the pain with narcotics, restoring blood volume, and withholding food (but giving parenteral hyperalimentation) in order to "rest" the pancreas and minimize the secretion of its enzymes.

Liver Disorders: Hepatitis and Cirrhosis

Disorders of the liver produce extremely serious complications, including jaundice, portal hypertension, hepatic encephalopathy, and ascites. Two common hepatic disorders are hepatitis and cirrhosis.

Hepatitis, inflammation of the liver, is associated chiefly with viruses called the *hepatitis A, B, C, D,* and *E viruses,* abbreviated HAV, HBV, etc. It can also result from alcohol abuse, certain drugs, and other diseases, but only viral hepatitis is discussed here.

The viral strains causing hepatitis differ in the following respects:

- **transmission**—through food or water contaminated with feces, parenteral means such as blood transfusion and contaminated needles, and sexual transmission;
- **prophylaxis (prevention)**—by vaccination, water purification, hygiene, and blood screening;
- **onset of symptoms**—acute (abrupt) or insidious (gradual);
- **severity**—for example, HAV produces mild symptoms, while the symptoms of other types are more severe, with HEV especially so in pregnant women; and
- **affected age groups**—that is, HAV and HEV affect especially children and young adults, and the others affect all ages.

HAV produces a mild infection and is very common. About 45% of people in urban areas show evidence (serum antibodies) of having had it. HAV spreads rapidly in settings such as day-care centers and institutions for mental patients. Infection is followed by long-lasting immunity. HBV is transmitted sexually and through blood and other body fluids. HCV is responsible for most post-transfusion hepatitis, but is becoming increasingly common as a sexually transmitted disease (already more prevalent than AIDS). HDV occurs only in people with HBV, because it depends on HBV for its own replication. HEV is commonly spread through contaminated drinking water in developing countries.

The disease progresses through three phases: prodromal, icteric, and recovery. The **prodromal phase** begins about 2 weeks after exposure and lasts 1 to 2 weeks. Signs and symptoms include fatigue, malaise, nausea and vomiting, and pain in the right upper quadrant of the abdomen. A person may become nauseated by the odor of food, and changes in the sense of taste inhibit the desire for tobacco or alcohol. A weight loss of 2 to 4 kg (4–9 lb) is common.

The transition from the prodromal to the **icteric phase** is marked by the appearance of jaundice. This phase lasts for the next 2 to 6 weeks. **Icterus** (jaundice) results from cellular destruction in the liver and the blockage of bile secretion, causing bile pigments to accumulate. Bile pigments are responsible for the brown color of feces. Consequently, the feces in an icteric patient are often gray or clay-colored, whereas the urine is abnormally dark. The liver is enlarged and tender, and signs of liver dysfunction appear, such as a rise in serum enzymes (aspartate transaminase and alanine transaminase) and prolonged blood clotting.

The **recovery phase** begins when the jaundice disappears and usually lasts 6 to 8 weeks. The symptoms diminish and liver functions return to normal. In HBV and HCV infection, however, chronic hepatitis may follow and last up to 6 months. Chronic hepatitis is a risk factor for cirrhosis and liver cancer, and HCV is the leading cause of liver failure and the need for liver transplants.

Hepatitis is diagnosed from the patient history, signs and symptoms, elevated serum enzyme levels, and detection of the hepatitis antigens. There is no specific treatment for acute hepatitis except to rest, eat a low-fat, high-carbohydrate diet, and wait for recovery. Interferon is sometimes useful in treating chronic hepatitis. Health-care providers can protect themselves from hepatitis by vaccination, avoiding contact with patients' blood and body fluids, and wearing gloves, especially if changing bedpans or being otherwise exposed to patients' wastes.

Cirrhosis is an irreversible inflammatory liver disease. It develops slowly over a period of years, but has a high rate of mortality, ranking as one of the leading causes of death in the United States. Cirrhosis is characterized by a disorganized liver histology in which regions of scar tissue alternate with irregular nodules of regenerating cells, giving the liver a lumpy or knobby appearance and a hardened texture. The bile passages become obstructed, leading to jaundice. Obstruction of the hepatic circulation stimulates *angiogenesis,* the growth of new blood vessels to bypass the congested liver. As blood bypasses the liver, the condition of the liver grows worse, with further necrosis and often, liver failure. The most common form of cirrhosis is **alcoholic cirrhosis**. Another type, **biliary cirrhosis,** has a *primary* form that has no known cause (possibly autoimmune) and a *secondary* form that stems from gallstones, chronic pancreatitis, or other conditions that obstruct the bile passages. A third type, **postnecrotic cirrhosis,** often follows chronic hepatitis C and other liver diseases.

Cirrhosis has several of the same signs as hepatitis—jaundice, pain, anorexia, weight loss, elevated hepatic enzymes in the blood serum, and reduced liver function (prolonged blood clotting and low blood albumin levels). Diagnosis is based on the patient history, risk factors, physical examination, hematology, and sometimes liver biopsy. The damage is irreversible, and the prognosis is often poor. However, treatment is possible and includes nutritional support, rest, control of gastrointestinal bleeding and ascites, avoidance of causative agents such as alcohol, use of immunosuppressive and anti-inflammatory drugs, and sometimes a liver transplant.

Inflammatory Bowel Diseases

Inflammatory bowel diseases are usually chronic and may affect the small intestine, the large intestine, or both. The three most common are ulcerative colitis, Crohn disease, and diverticulitis.

Ulcerative colitis is a chronic inflammatory disease of the large intestine, especially the rectum and sigmoid colon. It usually strikes between the ages of 20 and 40. The cause is unknown, but heredity is one risk factor since the condition is particularly common among people of Jewish descent, is more common among whites in general than among other groups, and occurs in identical twins more often than in nontwin siblings. The colonic mucosa becomes dark red, velvety, and swollen. Small lesions begin in the intestinal crypts and coalesce to form larger ulcers. The disease progresses to painful cramps, watery diarrhea, and the urgent need to defecate. In severe cases, a patient may have as many as 10 or 20 watery, bloody bowel movements per day. Edema and constriction of the colon can cause intestinal obstruction, and in some cases perforation, hypotension, and shock may occur. Ulcerative colitis is diagnosed by patient history, physical examination, hematology, sigmoidoscopy, and barium-enema X ray. It is treated with anti-inflammatory and antibiotic drugs, I.V. fluid replacement, and sometimes **colostomy** or surgical **resection** of the inflamed region.

Crohn disease is similar to ulcerative colitis in pathology, and the two are sometimes difficult to differentiate. Crohn disease, however, affects both the large and small intestines and seldom involves the rectum. The ileocecal region is the most commonly affected. The two diseases are similar in risk factors and etiology. Crohn disease produces fissures that cross the intestinal wall and may involve both mucosa and serosa. These fissures and the areas of edema between them give the intestinal wall a cobblestone appearance. The chief symptom is an "irritable bowel" that may last for several years, producing diarrhea, abdominal tenderness or pain, and weight loss. Malnutrition may result from malabsorption of vitamin B_{12}, vitamin D, and calcium. Crohn disease may be treated with immunosuppressive drugs and, if necessary, with surgery for resulting obstructions, abscesses, or **fistulae** (abnormal passages).

Diverticulitis is the inflammation of *diverticula*—saclike herniations of the mucosa of the colon through the muscular wall. It is most common in elderly people in developed countries where the diet has a high proportion of refined foods and a low fiber content. The most commonly affected region of the GI tract is the sigmoid colon. The condition is worsened by low-fiber diets because the lack of bulk in the stool reduces the diameter of the colon, increasing both the pressure in the lumen and the risk of herniation. Diverticulitis causes painful cramping, diarrhea or constipation, and flatulence. Occasionally, a diverticulum ruptures, and **peritonitis** (inflammation of the peritoneum) may ensue. Diverticulitis is diagnosed in the same way as ulcerative colitis. The symptoms are sometimes relieved by increasing the amount of fiber in the diet, but severe cases can require surgical resection.

Appendicitis

Appendicitis is inflammation of the vermiform appendix. It occurs in 7% to 10% of the population and most often strikes between the ages of 20 and 30. The cause of appendicitis is still uncertain. The most popular theory is that it begins with obstruction of the appendix by compacted feces, a tumor, or a foreign body that a person has swallowed. The obstruction blocks drainage of the appendix, so fluid accumulates, pressure increases, and the compression of blood vessels in the appendix leads to hypoxia, necrosis, and ulceration. This progresses to bacterial invasion, thrombosis, gangrene, and sometimes perforation. If the appendix is perforated, the spread of bacteria into the peritoneal cavity causes peritonitis.

The signs of appendicitis begin with a vague pain in the epigastric or umbilical region. The pain becomes more intense over the next 3 to 4 hours, then may subside and return later in a new location, the right lower quadrant (RLQ) of the abdomen. The pain is accompanied by fever, nausea, and vomiting.

Physical examination typically shows rebound tenderness in the RLQ; hematologic work shows both the neutrophil count and total WBC count to be elevated; and the appendicitis may be confirmed by ultrasound, a CT scan, or **laparoscopy** (viewing the abdominal cavity with an endoscope inserted through a small umbilical incision).

The treatment is **appendectomy,** the most common emergency surgery of the abdomen. Major appendectomy scars are now largely a thing of the past since laparoscopic surgery allows the procedure to be performed through just three small incisions, one of which is concealed in the umbilicus. Appendectomy is often performed even upon suspicion of appendicitis because it is too risky to wait. However, because of this, about one appendicitis surgery in six shows the appendix to be healthy, leaving the true source of the pain still to be identified. In this event, surgeons examine the ileocecal area for signs of inflammatory bowel disease and examine female patients for ovarian cysts, *salpingitis* (inflammation of the uterine tubes), or *ectopic pregnancy* (implantation of an embryo in the uterine tube).

Colorectal Cancer

Colorectal cancer (cancer of the lower intestinal tract) causes 10% to 15% of all cancer-related deaths annually in the United States. Worldwide, its incidence is increasing, especially in populations where the diet tends to be low in fiber and high in animal protein, fat, and refined carbohydrates. Aside from diet, the risk factors for colorectal cancer include aging (most cases occur after age 50, with peak incidence between the ages of 60 and 75), diverticulitis, chronic ulcerative colitis, and a family history of colorectal cancer. The development of colorectal cancer has been linked to mutations in tumor suppressor genes called *p53, APC,* and *DCC* and activation of the *ras* oncogene.

Patients in the early stages of colorectal cancer do not usually exhibit specific signs and symptoms and are therefore less likely to seek help right away. The patient may complain of lower abdominal pain and rectal bleeding. One of the best diagnostic tools is an occult blood test, in which a small fecal sample is spread on a card with a chemical reagent that will disclose the presence of even invisible quantities of blood. If colorectal cancer is suspected, imaging tests confirm the diagnosis.

Treatment consists of surgically removing the cancerous region and associated lymphatic components. Radiation therapy may be used prior to surgery to decrease the size of the tumor. Following surgery, chemotherapy and immunotherapy may be employed to treat metastasized tumors and limit the recurrence of the cancer. The 5-year survival rate following surgery is greater than 90% for cancer limited to the mucosa. As metastasis increases, survival rates decrease proportionately.

Case Study 25 The Man with Yellow Eyes

Walter works as a salesman for a software development firm. In connection with his job, he frequently travels to countries throughout the world. Walter goes on a sales trip to Southeast Asia, and after his business is finished, spends 2 extra weeks touring the ancient ruins of Thailand and sampling the night life in Bangkok. About 3 months after his return, Walter notices that he is constantly tired, has a slight fever, and often feels nauseous. Although Walter normally enjoys a good dinner with cocktails and is a smoker, he now has little interest in eating and finds cigarettes and alcohol distasteful.

After a few days of these symptoms, Walter thinks that he is improving and returns to his normal activities. But the next week he notices that his urine has turned a dark color and his abdomen is slightly tender. Fearing that he may have contracted a disease on his trip, Walter makes an appointment to see his physician the next day.

The physician completes a physical examination and notes that Walter's corneas and oral mucosa are slightly yellowed. Since Walter is slightly tanned from his travels, this jaundice is not as apparent in his skin, but is present there as well. In completing the patient history, the physician asks Walter about his immunization record and his activities during his travels. Walter is current for most immunizations (mumps, measles, diphtheria, polio). He says that he thought of getting vaccinated for hepatitis before his trip, but decided against it because he did not have time to complete the series. Walter mentions that he is always careful to drink bottled water during his travels. Afraid that he might have AIDS, Walter finally confides that he has had sexual relations with both men and women in his home territory and on his stay in Asia, although he says that all of his partners appeared to be healthy.

Results of the physical examination and blood tests are shown here.

Vital signs:

Oral temperature = 98.7°F (37.1°C)
Heart rate = 74 beats/min
Respiratory rate = 13 breaths/min
Blood pressure = 136/76 mmHg

Physical examination:

Skin and mucous membranes: Yellow tinge but otherwise normal.
Musculoskeletal: No abnormalities noted.
Nervous and sensory: No abnormalities noted.
Cardiovascular: No abnormalities noted.
Respiratory: No abnormalities noted.
Abdominal organs: Hepatomegaly; slight splenomegaly; normal bowel sounds; hepatic tenderness and pain upon percussion and palpation.

Blood:

Serum aspartate transaminase = 595 IU/L
Serum alanine transaminase = 1,948 IU/L
Serum bilirubin = 1.1 mg/dL
Negative for HIV antigens
Positive for hepatitis B antigens
All other findings normal

Based on these results, Walter is diagnosed with hepatitis B. His physician tells him the disease will most likely run its course and he will recover within a few weeks. In the meantime, he advises Walter to limit his physical activity until his appetite returns and to go back to work only after the jaundice resolves. He points out that a follow-up examination is necessary because of the slight possibility of the infection developing into chronic hepatitis, which would require further treatment.

The physician also suggests that Walter -consider following a low-fat, high-carbohydrate diet, although this is not essential to his recovery. Finally, he counsels Walter about the importance of using condoms and controlling his sexual activity to minimize the chance of contracting sexually transmitted diseases in the future.

Based on this case study and other information in this chapter, answer the following questions.

1. What factors in Walter's history suggest a tentative diagnosis of hepatitis?
2. What laboratory findings confirm this diagnosis?
3. If the laboratory hematologic findings had not yet come in, but you suspected Walter had hepatitis anyway, why would you be unlikely to suspect hepatitis A?
4. If Walter were to develop chronic hepatitis, he would be at high risk of cirrhosis of the liver. What additional signs and symptoms would indicate that this was occurring?
5. One function of the liver is to produce most of the proteins of blood plasma. Of the following conditions associated with reduced liver function, which one would be caused by the reduced protein synthesis?
   a. ascites
   b. cirrhosis
   c. jaundice
   d. fatty liver
   e. weight loss
6. Jeremy, a 32-year-old man, visits his physician complaining of abdominal pain, frequent bowel movements, and bouts of diarrhea. Based on this information, which of the following diseases does Jeremy most likely have?
   a. Crohn disease
   b. hepatitis
   c. colorectal cancer
   d. pancreatitis
   e. appendicitis
7. Pancreatitis is treated partly by withholding food orally and replacing it with parenteral hyperalimentation. Explain why.
8. Identify and describe a positive feedback loop that occurs in cirrhosis with angiogenesis.
9. Why does cirrhosis of the liver cause poor fat digestion and reduced absorption of vitamins A, D, E, and K? Why does it not have a similar effect on absorption of vitamin C?
10. Ulcerative colitis and Crohn disease have very similar signs and symptoms, but only Crohn disease is likely to produce malnutrition. Explain why.

Selected Clinical Terms

abdominal guarding Reflex contraction of the abdominal muscles occurring upon palpation to minimize motion of areas made tender or painful by disease; often indicates peritonitis.

anorexia Loss of appetite; aversion to food.

colostomy Surgical creation of a new opening from the abdominal wall into the colon, necessitated by the removal of diseased portions of the colon.

dyspepsia Epigastric pain, burning, nausea, or gas resulting from a disorder of the stomach; "upset stomach" or "indigestion."

dysphagia Difficulty swallowing.

fistula Any abnormal passage between two epithelium-lined spaces or surfaces, such as a pathological opening between the rectum and bladder or between the esophagus and a bronchus.

icterus Synonym for jaundice.

jaundice Abnormal yellow coloration of the skin, conjunctiva, or mucous membranes due to accumulated bile pigments; often indicates hemolytic anemia or liver dysfunction.

laparoscopy Endoscopy of the abdominal cavity, typically by inserting an endoscope through a small incision in the umbilicus.

occult blood Blood in the stool or elsewhere that is too scanty to be visible but can be detected chemically; may indicate colon polyps, cancer, hemorrhoids, or other disorders.

peritonitis Inflammation of the peritoneum, often caused by perforation of the appendix or rupture of a diverticulum in diverticulitis.

rebound tenderness A patient's sensation of pain felt not while the examiner palpates the body surface but when the hand is removed and pressure is released.

resection Surgical removal of part of an organ, such as a segment of colon affected by cancer or ulcerative colitis.

steatorrhea The presence of an abnormal amount of fat in the stool, indicating poor fat digestion or malabsorption; typical of diseases that reduce bile secretion.

26 Nutrition and Metabolism

Objectives

In this chapter we will study

- procedures used for nutritional assessment of a patient;
- methods of nutritional support;
- varieties, pathology, and treatment of malnutrition;
- two common eating disorders—anorexia nervosa and bulimia nervosa; and
- fad diets and how to recognize suspicious dietary advice.

Clinical Approaches to Nutrition

Understanding and appreciating nutrition has become increasingly important in our everyday lives. We are constantly bombarded with advertisements and books promising rapid and safe weight loss. On the other hand, prolonged illnesses and certain medical procedures can induce *malnutrition,* a lack of one or more nutrients due to dietary deficiency or the inability to properly absorb nutrients. Inherited disorders affecting metabolism, such as phenylketonuria and cystic fibrosis, require specific diets for proper management. In developing countries or other areas with limited economic means, states of malnutrition such as *kwashiorkor* and *marasmus* are relatively common. And just as a lack of certain vitamins and minerals can cause disease, too much of a certain substance, as in hypervitaminosis, can also have harmful effects.

Interactions between nutrients and drugs are common. Malnutrition can impair drug absorption or metabolism, leading to potentially hazardous side effects. Drugs may also alter appetite or interfere with the metabolism of specific macromolecules, leading to energy and nutrient imbalances. Finally, drugs may hinder the absorption of vitamins and minerals.

Assessing Nutritional Status

A nutritional assessment determines a patient's health from a nutritional perspective. An abbreviated assessment can be conducted by a clinician, while a complete assessment is most often performed by either a dietitian or a clinician who has advanced training in nutrition. Either type employs many of the same procedures involved in obtaining a patient history and conducting a physical examination.

A complete nutritional assessment consists of the patient's history, body measurements, a physical examination, and laboratory tests (if warranted). The patient history provides information about the risk factors for poor nutrition, which can be divided into four broad categories:

1. The **health history** identifies specific diseases that predispose a patient to poor nutrition, such as AIDS, cancer, diabetes mellitus, lung disease, or physiological conditions such as obesity, low body weight, or anorexia.
2. **Socioeconomic status** indicates whether the patient is at risk for nutritional deficiency. Such factors as having no one to eat with, little money for food, inadequate food storage or preparation facilities, minimal education, poor self-esteem, and lack of transportation all predispose a patient to nutritional deficit.
3. **Medications** can also increase the risk of nutritional imbalance. These include analgesics, antacids, antidepressants, diuretics, immunosuppressants, laxatives, and oral contraceptives.
4. **Dietary history,** including known eating disorders, poor appetite, or a history of dieting, also provides useful background information.

After the dietary history is completed, various physical measurements are taken to determine body composition and development. These include the following:

- **Height, weight, and (in children) head circumference** The ratio of height to weight is used to determine whether a patient is overweight or underweight. The relationship of head circumference to height and weight is a measure of whether a child is growing at a normal rate.

- **Midarm circumference** This is used as an indicator of muscle mass and protein balance. The circumference of the arm is measured around the thickest part of the biceps, and then the thickness of a fat fold over the triceps is measured with calipers. A standard formula is used to calculate midarm muscle mass from these data.
- **Waist-to-hip ratio** A measurement of the ratio of waist circumference to hip circumference helps determine whether a patient has a normal fat distribution. A waist-to-hip ratio exceeding 0.8 in women and 0.95 in men indicates increased health risk due to obesity. For example, if a person has a 30-inch waist and a 32-inch hip measurement, the waist circumference would be divided by the hip circumference, yielding a waist-to-hip ratio of 0.94. For a woman, this would indicate an increased risk for disease, but for a man it would be considered in normal range.
- **Fat fold measurements** Fat folds are measured in areas such as the chin, biceps, triceps, thigh, and umbilicus to determine a patient's percentage of body fat. This is a more accurate way to determine obesity than by simply comparing the patient's weight and height.

Vitamin, mineral, and protein deficiencies can often be recognized by skillful examination of the patient's external appearance. For example, abnormalities of bone shapes, skin and hair texture, nail shapes and markings, teeth, tongue, and muscle tone can be signs of specific nutrient deficiencies. An array of blood and urine tests are also available to detect deficiencies in specific nutrients. Dietitians and clinical nutritionists usually work with a physician or health-care team and can request such clinical tests.

An abbreviated nutritional assessment includes measuring body weight, serum lymphocytes, albumin and transferrin, triceps skinfold thickness, upper-arm circumference, and the skin *anergy* (impaired or absent response) to a number of antigens. Abnormal values serve as a measure of the degree of nutritional deficit. This information is especially important when determining the risk of complications in a patient slated for surgery.

Providing Nutritional Support

Nutritional support is classified as either *enteral* or *parenteral.* In **enteral nutrition,** a nutritional supplement is given to the patient either orally or through a "feeding tube." *Oral enteral nutrition* is used when the patient has normal digestive system function but requires supplementation with calories or specific nutrients. Patients with chronic inflammatory disease or cancer may be placed on such diets.

Enteral tube alimentation uses a nasogastric or nasoduodenal tube placed into the nose and down the esophagus to deliver nutrients directly into the duodenum or just proximal to it. This procedure is used for patients who have functional digestive systems but need nutritional supplementation because of conditions such as trauma to the head or neck, coma, burns, anorexia, or severe malnutrition. It is also used to totally replace oral feeding in patients who refuse to eat. The supplementation may be a high-energy, high-protein, or chemically defined diet consisting of essential nutrients that can be readily absorbed (to minimize fecal volume). The rate of feeding is carefully monitored to minimize diarrhea and GI discomfort. The most common side effects of enteral tube alimentation are electrolyte imbalances, hyperosmolarity, and volume overload.

Parenteral nutrition provides either partial or total nutrition intravenously. It is called *partial parenteral nutrition* when certain nutrients are given intravenously to supplement oral intake and *total parenteral nutrition (TPN)* when all of the patient's nutrition is given this way. In cases where GI function has been permanently lost, parenteral nutrition is continued at home *(HPN)* after the patient is discharged. Parenteral nutrition entails several potential complications, including thrombosis at the site of I.V. administration, air embolism, volume overload, metabolic disturbances such as hyperglycemia and electrolyte imbalances, and infection (the most common complication).

Undernutrition

Malnutrition is any form of improper nutrition resulting from any cause: undereating, overeating, an imbalanced diet, or disturbances in the body's ability to absorb and use specific nutrients. Here we discuss one side of the issue, **undernutrition,** which is malnutrition that results from inadequate food intake or defects in the digestion, absorption, and use of nutrients. Undernutrition progresses through four stages: (1) reduced nutrient levels in the blood and other tissues, (2) changes in cellular enzyme activity due to lack of nutrients, (3) tissue and organ dysfunction, and (4) outward signs of undernutrition such as weight loss and lethargy. The most recognizable forms of undernutrition are *starvation* and *protein-energy malnutrition.*

Starvation

Starvation is the prolonged deprivation of all food. The body's first physiological priority is to maintain the blood glucose level to provide adequate energy to the brain, which is unable to employ any alternative energy source. Liver glycogen is broken down *(glycogenolysis)* and its glucose released into circulation. Glycogen stores are depleted in just a few hours, and the body then resorts to *gluconeogenesis,* the conversion of other organic fuels into glucose. Gluconeogenesis is stimulated chiefly by cortisol. Fat is broken down first, and at this stage starvation is reflected in noticeable weight loss. When fat stores are gone, the body begins to break down protein. This results in **inanition,** the general wasting away, weakness, depressed metabolism, and debility of the body as a person loses muscle mass and strength. This is usually seen after weeks to months of starvation. (Inanition can also result from such conditions as **malabsorption,** cancer, and AIDS.)

Aside from inanition, some signs of starvation include dry, thin, inelastic skin that may have patchy, brown pigmentation; pallor; dry, brittle hair and alopecia (thinning of the hair); diarrhea; hypotension; bradycardia and reduced cardiac output; anemia; depressed immune responses; gonadal atrophy; and hypothermia. Hematologic tests show elevated plasma amino acid and lipid concentrations coupled with a low level of insulin. With no food intake whatsoever, starvation is fatal in 8 to 12 weeks (assuming a person doesn't first die of dehydration).

The treatment of starvation depends on its severity. In extreme cases, the GI tract may be so atrophied that total parenteral nutrition is required. Nutrients must be restored gradually. In many cases, the first oral foods are liquids such as juice or broth. After a few days, a solid diet is started. Small, frequent meals are advisable to avoid diarrhea. Intake equal to or in excess of 5,000 Calories a day is recommended, but weight gain should not exceed 2 kg (4 lb) per week.

Protein-Energy Malnutrition

Protein-energy malnutrition (PEM) is an inadequate intake of protein, calories, or both, typically seen in children under age 5 and usually in impoverished countries or in famines. The World Health Organization estimates that several hundred million children worldwide have PEM, with disastrous consequences in terms of childhood mortality, children's mental development, and a nation's social organization and economic development. If 100 jumbo jets crashed every day, each carrying 400 young children to their deaths, it would cause a global uproar; yet that many children die every day, quietly and unnoticed by most, from PEM and associated conditions. PEM is also common in AIDS patients.

The two most common manifestations of PEM are as follows:

- **Marasmus** results from a dietary deficiency of both calories and protein. This forces the body to burn what dietary protein it does get for fuel rather than using it for tissue growth and maintenance. Growth is stunted and adipose tissue and muscle (protein) mass are lost, but edema does not occur (compare the next syndrome).
- **Kwashiorkor** results from a selective deficiency of protein, as when too much of the caloric intake comes from nonprotein sources such as rice or other plant foods. This can also occur in patients who cannot eat and are given only I.V. glucose. As with marasmus, kwashiorkor stunts childhood growth, but subcutaneous fat is preserved and the liver is fatty. The blood plasma is low in protein *(hypoproteinemia),* which upsets the balance between capillary filtration and reabsorption and thus leads to edema and ascites.

Additional effects of PEM include anemia; reduced blood volume, reproductive function, and metabolic rate; atrophy of the digestive tract; hypothermia; reduced immune responses; and poor wound healing. Cardiac, pulmonary, and renal function decline but remain proportional to body mass and metabolic state. However, aggressive treatment to restore nutrition and fluid volume can cause the diminished organ systems to fail. Treatment of severe PEM therefore requires careful and gradual restoration of nutritional status and fluid volume. Gastrointestinal atrophy may make it impossible for a patient to take enteral nutrition immediately; total parenteral nutrition may be necessary. Vitamin and mineral supplementation is especially important.

Eating Disorders

While people in some parts of the world suffer from lack of food, those living in areas where food is plentiful also have nutritional deficiencies. Some of this malnutrition results from such eating disorders as anorexia nervosa and bulimia nervosa. It has been estimated that upwards of 2 million people in the United States suffer from these two disorders. Both of them occur overwhelmingly in young, white, middle-class women obsessed with becoming or staying thin. They are quite rare among women of African and Asian descent and the poor, and have about one-tenth the incidence in men as in women.

Anorexia Nervosa

Anorexia nervosa is the radical, self-imposed restriction of food intake for the purpose of staying thin. It typically leads to emaciation. Women with anorexia nervosa tend to tie their self-esteem to their thinness and to deny the seriousness of their low weight. Anorexia nervosa is diagnosed from a history of major weight loss in the absence of organic disease, with a *body mass index (BMI)* of 18 or less.

The cause of anorexia nervosa is not completely known. It seems typically to stem from a complex web of interpersonal relationships in extremely achievement-oriented families; from social norms that attach undue importance to slimness; and from pursuits such as dance and athletics (especially running). Many women with anorexia nervosa were overweight in childhood. Anorexia typically begins shortly after puberty and usually not later than the middle 20s. It is especially common among college women.

Some signs of anorexia nervosa include amenorrhea (missing three or more consecutive menstrual periods), constipation, cold intolerance, hypotension, hypokalemia, azotemia, and bradycardia. Body fat can become virtually undetectable (although breast fullness tends to be preserved), and the bones protrude conspicuously. However, these signs may not be apparent in a dressed patient because edema of the legs and enlargement of the parotid glands can make the legs and face appear normal. The skin is often dry, scaly, and somewhat yellow. Women with anorexia nervosa refuse to maintain at least 85% of their ideal body weight. When weight falls 35% below the ideal, a patient is at high risk of sudden death from cardiac arrhythmia. The incidence of suicide is also high.

Treatment involves both nutritional therapy and psychiatric counseling, but anorexic patients are often highly resistant to therapy. Serious cases require hospitalization and realimentation. Long-term follow-up studies of anorexia patients show that about half achieve normal weight, 20% improve but remain underweight, 20% continue to be anorexic, 5% become obese, and 6% die.

Bulimia Nervosa

Bulimia nervosa is similar to anorexia nervosa in its psychological cause, but different in habit and effect. Bulimia is diagnosed when the patient binges (eats excessive amounts of food) at least twice a week for 3 months, feels his or her eating is out of control, and takes measures to compensate for it and keep the weight down, such as inducing vomiting or using laxatives to purge the digestive system. Some bulimics follow bingeing with frenzied exercise. *Bulimia* literally means "ox-hunger." For all their efforts, bulimics usually do not lose a significant amount of weight. Antisocial and obsessive/compulsive behavior and self-mutilation are common in bulimia. Most bulimics binge at least once a day, with a binge lasting an average of 1.2 hours. Bingers consume as much as 10,000 kilocalories at one sitting, most often in such forms as ice cream, bread, candy, cookies, potato chips, cakes, doughnuts, and soft drinks. Bulimics are more secretive about their eating behavior than anorexics, so family and friends are often less aware of the disorder. Depression and suicide are more common in bulimia than in anorexia.

Some signs of bulimia nervosa include fluctuating weight, scars from self-mutilation, pharyngitis, and frequent urinary tract infections. Frequent self-induced vomiting leads to hypokalemia, and the stomach acid inflames the esophagus and erodes the tooth enamel. Bulimics are at risk of suffocation on aspirated vomit, rupture of the stomach or esophagus, pancreatitis, and cardiac arrhythmia. Bulimics have a poorer prognosis for recovery than anorexics. Up to 40% remain bulimic after 18 months of therapy, and about two-thirds of those who recover relapse within a year. Antidepressants are more helpful in treating bulimia than anorexia.

Dieting and Weight Loss

The number of advertisements and books describing diets for weight loss or muscle gain is staggering. Very few individuals seem satisfied with their body shape, and many want the "quick fix" offered by *fad diets*. Although such diets promise rapid results, many of their claims are false, unsubstantiated, or based on theories of weight loss lacking scientific merit. They are designed to sell books and nutritional products, not to help the buyer. Some fad diets are relatively harmless, but others can produce serious side effects ranging from headaches, dizziness, and nausea to cardiac arrhythmia, ketoacidosis, coma, and death. It has been estimated that of the almost 30,000 weight-loss theories, treatments, and claims, less than 6% provide long-term benefits.

How can you recognize a fad diet? It is often easier than you might think, since they share a number of general characteristics:

- The diet claims to produce rapid weight loss of more than 1% of total body weight per week. Although this loss may actually occur, it is often due mainly to the loss of water, glycogen, and muscle.
- The plan dictates that the dieter consume only certain foods (such as nothing but grapefruit for breakfast or no carbohydrates) or that calories be drastically restricted (such as no more than 800 Calories (kilocalories) per day).
- The seller claims that the diet works for everyone, recognizing no distinctions between different people or types of weight control problems.
- The diet requires the user to eat only food that is provided by the company promoting it, often at high prices.
- The diet claims to have a "special" or "magical" ingredient.
- The promoters often try to cast suspicion on scientists, dietitians, and physicians, encouraging dieters to take advice from outside the medical establishment.
- Celebrities or past clients provide testimonials to the diet's success, and the plan is often linked to glamorous cities such as Beverly Hills and New York.

In all of these instances, little if any attention is paid to helping the dieter modify his or her diet and eating behavior. What often happens is that the dieter does indeed experience weight loss, but once he or she stops dieting, the lost weight (and often more) is regained. This cycle of weight loss and gain is known as *weight cycling* and may actually cause people to gain more weight in the long run than if they had not dieted at all.

Effective weight-loss programs emphasize modifying the diet to achieve slow, steady weight loss by decreasing total caloric intake while maintaining a balance of nutrients and reducing the number of calories consumed as fat. Most responsible plans advocate consuming no less than 1,200 Calories per day. They also recommend behavior modifications such as limiting snacking, not skipping meals, eating at set times and in set locations, and exercising to increase caloric expenditure and muscle mass.

Case Study 26 The Wrestler Who Wanted to Make Weight

Marcus is a member of his high school wrestling team. Recently, he has been gaining weight, and he is concerned that he might be moved up to the next weight class, where he would be less competitive. Marcus has been trying a variety of methods to keep his weight down while preserving muscle, including running, taking saunas, fasting for a day before weigh-in, and eating a high-protein diet recommended by his coach. In spite of his efforts, Marcus's hunger is never satisfied—he eats more than the diet recommends and continues to gain weight. He decides to try the diet pills that his mother and sister use, and to eat enough to feel satisfied but then induce vomiting.

This seems to work for a few weeks, but then Marcus finds that the diet pills are not suppressing his appetite enough. He stops taking them, but eats more before vomiting. At the end of the season, Marcus's coach encourages him to maintain his weight over the summer. Marcus therefore continues to binge and purge, doing it secretly so his parents won't know. However, during a routine dental checkup that summer, the dentist notes some erosion of Marcus's tooth enamel and enlargement of his salivary glands. Recognizing the signs of bulimia nervosa, the dentist mentions this to Marcus and his father, and Marcus admits that he has been inducing vomiting to control his weight. The dentist recommends that Marcus see the family physician for a physical. The physician observes slight dehydration and esophageal inflammation, and notes that Marcus's weight is only 70% of the desirable weight for his height. He confirms bulimia nervosa, advises Marcus and his father on its serious complications, and recommends psychological counseling. However, Marcus convinces his father that he doesn't need to see a psychologist and will change his behavior.

Marcus stops bingeing and vomiting, but after gaining 10 pounds, he decides to try laxatives to lose weight, as some of his teammates do. He thinks he can keep his weight down this way without risking the health consequences his doctor described.

One day in practice, Marcus breaks his ulna. The radiologist at the hospital notes that Marcus's bone density is low for his age and recommends that he see his family doctor. The family physician finds that Marcus still weighs only 75% of the ideal for his age and height. A bone densitometry test shows that he has osteomalacia. Laboratory tests show that he also has anemia, hypokalemia, hypovolemia, and hypoproteinemia. In addition, Marcus's ECG shows cardiac arrhythmia. Marcus admits that he has been bingeing and purging again. He is hospitalized for realimentation and begins counseling with a psychologist and nutritionist for bulimia nervosa.

Based on this case study and other information in this chapter, answer the following questions.

1. Why is Marcus diagnosed with bulimia nervosa rather than anorexia nervosa?
2. Do you think Marcus's realimentation program would be administered enterally or parenterally? Why?
3. Why might Marcus be placed on a realimentation diet that is especially high in protein and vitamin D? What else do you expect might be included in his diet? Why?
4. Explain why Marcus is dehydrated. Explain why he is hypokalemic.
5. Suppose Marcus had never tried anything except diet pills and laxatives. How would his signs and symptoms differ? What signs and symptoms would remain the same?
6. Explain why Marcus develops osteomalacia.
7. *Staphylococcus aureus* infection is a major concern for patients on home parenteral nutrition. Explain why.
8. Explain why an elderly, bedridden patient undergoing enteral alimentation through a nasoduodenal tube could be especially susceptible to volume overload and hyperosmolarity.
9. A television commercial touts the wonders of a new nonprescription diet pill said to control weight gain by binding fat and preventing it from being digested and absorbed. Assuming the pill does what it claims, describe as many undesirable side effects as you can think of that the pill could have.
10. American soldiers who liberated the Nazi concentration camps at the end of World War II were so moved by the inmates' emaciation that they offered them candy bars and other rich food. In spite of the inmates' extreme starvation, these had to be taken away from them. Explain why.

Selected Clinical Terms

anorexia nervosa Severe, self-imposed restriction of food intake due to an exaggerated fear of becoming fat; leads to emaciation, amenorrhea, and other consequences, including the possibility of eventual heart failure.

bulimia nervosa A habit of bingeing on enormous quantities of food and then attempting to compensate in an exaggerated effort to keep from gaining weight, usually by inducing vomiting or using laxatives, but without substantial weight loss.

enteral nutrition Nutrition taken by way of the alimentary canal, either orally or by nasogastric tube, and requiring intestinal absorption.

inanition Body wasting, weakness, depressed metabolism, and debility resulting from starvation or certain chronic diseases.

kwashiorkor A state of stunted growth, muscle loss, hypoproteinemia, edema, and ascites resulting from prolonged protein deficiency, but not necessarily accompanied by caloric deficiency.

malabsorption A defect in nutrient absorption by the digestive tract, so that malnutrition can occur in spite of adequate nutrient intake.

malnutrition Improper nutrition due to overeating, undereating, an imbalanced diet, or malabsorption of nutrients.

marasmus A state of stunted growth and loss of fat and muscle due to prolonged protein-energy malnutrition.

parenteral nutrition Nutrition taken by routes other than the alimentary canal, usually intravenously, in cases where a patient cannot swallow or the condition of the digestive tract is unfavorable to normal digestion and absorption.

protein-energy malnutrition (PEM) Severe deprivation of protein, calories, or both.

starvation Prolonged deprivation of all food.

undernutrition A form of malnutrition due to inadequate food intake or to defects in nutrient digestion, absorption, or assimilation.

27 The Male Reproductive System

Objectives

In this chapter we will study

- the common symptoms of disorders of the male reproductive system;
- diagnostic procedures for examining the male reproductive system;
- the causes and course of gynecomastia;
- the scrotal disorders hydrocele, varicocele, and spermatocele;
- several testicular disorders—ectopic testis, torsion of the testis, and testicular cancer; and
- prostatic disorders, specifically prostatitis, benign prostatic hyperplasia, and prostate cancer.

Assessment of the Male Reproductive System

Because the male reproductive system is closely related to the urinary system, many of the same diagnostic procedures are used to evaluate both. Additional physical evaluations and laboratory tests are available for assessing the reproductive system alone.

Pain is one of the most common symptoms of male reproductive system disease. It may be diffuse pain in the groin or pain localized to a specific area, most often one of the testes. Since the penile urethra is the passageway for both urine and semen, painful urination may indicate either a urinary or a reproductive system disorder. Males may also experience discharge from the urethra. This is one of the most frequent symptoms of a sexually transmitted disease (STD), or it may arise from inflammation of the epididymis *(epididymitis)* or the prostate gland *(prostatitis)*.

Testicular pain may indicate an STD, testicular torsion, cancer, or cryptorchidism (failure of one or both testes to descend into the scrotum). Pain in the groin may be due to a muscle pull, an STD, or an inguinal hernia. Men also seek medical attention for infertility and erectile dysfunction.

Physical Examination

When obtaining a patient history in cases that involve the reproductive system, the clinician inquires about the patient's sexual history and practices as well as the onset of the symptoms relative to the most recent sexual contact. Because these are such personal questions, the clinician must assure the patient that honest answers are essential to an accurate diagnosis and that the information will remain confidential.

Examining the external genitalia and palpating the prostate gland are routine. Because many STDs produce characteristic lesions on the penis or scrotum, these structures are inspected for vesicles, chancres, or warts. The color and texture of the skin are also noted. Excessive redness may indicate inflammation, while dry, scaly skin may be a sign of fungal infection ("jock itch") or a nutrient imbalance. If the man has not been circumcised, the foreskin should be retracted and the glans examined. Inflammation of the prepuce or adjacent tissues prevents retraction of the foreskin, a condition called **phimosis.**

Following visual inspection, the scrotum, testis, epididymis, and ductus deferens are palpated on both sides, and a **digital rectal examination (DRE)** is conducted. Palpation allows the clinician to check for inflammation, edema, masses, or other abnormalities as well as to determine whether both testes are present and in the proper location. A DRE involves palpating the prostate and seminal vesicles through the anterior rectal wall by inserting a gloved finger into the rectum. The DRE is used to screen for prostate disease and inflammation of the seminal vesicles. If urethral discharge appears at any time during the physical examination, it is cultured for pathogenic microbes.

Laboratory Tests and Imaging Methods

If the patient history and physical examination indicate that further tests are advisable, a variety of diagnostic procedures are available. Blood samples are examined for reproductive hormones (testosterone, estradiol, luteinizing hormone, and follicle-stimulating hormone) and antibodies against organisms that cause STDs, such as *Treponema pallidum* and HIV. A high level of **prostate-specific antigen (PSA)** and alkaline phosphatase may suggest prostate cancer, while α-fetoprotein and human chorionic gonadotropin (HCG) point toward testicular cancer.

Semen analysis includes measures of semen volume and sperm count and assessment of sperm motility and morphology. The composition of the semen is also determined if abnormal function of one or more of the accessory glands is suspected. Either the semen or a urethral discharge may be cultured for the presence of infectious agents that cause various STDs. Finally, a biopsy may be done if cancer or abnormal structure and function at the cellular level is suspected.

X ray, CT, MRI, and sonography can be used to help diagnose cancer and congenital abnormalities of the reproductive organs. In sonography, the ultrasound transducer is inserted into the rectum so that the ultrasound waves travel only a short distance to the prostate and back, and a clearer image is produced. Sonography is also helpful in directing biopsy needles to the appropriate location, since it produces a real-time image of moving objects.

Gynecomastia

Gynecomastia, enlargement of the male breast, affects upwards of 40% of men in the United States and is seen most often in adolescents and men over the age of 50. It occurs when the ratio between testosterone and estrogen shifts in favor of estrogen, a phenomenon that tends to happen in men during puberty and aging. A variety of conditions can shift the testosterone/estrogen ratio. Hypogonadism, Klinefelter syndrome, cirrhosis and hepatitis, hyperthyroidism, tuberculosis, and cancers of the testicles, adrenal glands, and liver can cause gynecomastia. It can also result from drugs such as estrogen supplements, amphetamines, digitalis, spironolactone, ergotamine, and certain antidepressants.

Gynecomastia is readily diagnosed by physical examination. The breast stroma shows hyperplasia, producing a palpable mass (at least 2 cm in diameter) beneath the areola. Gynecomastia usually regresses spontaneously—the pubertal form within 4 to 6 months and the senescent form within 6 to 12 months. If it does not, or if an underlying treatable cause is known, that cause should be treated. Breast regression then usually follows within 12 months. If it does not, the man is instructed in breast self-examination since persistent gynecomastia can develop into breast cancer. Breast cancer occurs in about 1 male for every 275 female patients and accounts for about 0.2% of cancer cases among American males.

Scrotal Disorders

Some common noncancerous disorders of the scrotum are hydrocele, varicocele, and spermatocele. All can be detected by palpation.

Hydrocele, a collection of fluid in the tunica vaginalis, is the most common cause of scrotal swelling. It results from increased fluid production (often due to inflammation) or decreased reabsorption caused by either lymphatic or venous blockage. The underlying cause is usually trauma or infection of the testis or epididymis, but some cases are idiopathic. Hydroceles vary from pea-sized to bigger than a grapefruit. They can compress the testicular artery and vein, thus reducing testicular circulation and leading to atrophy. Hydrocele is treated by aspirating the accumulated fluid. If it recurs, a sclerosing drug is injected into the scrotum to induce scarring of the tunica vaginalis in an attempt to prevent recurrence. If this fails, the tunica vaginalis is surgically removed.

Varicocele, abnormal dilation of veins in the spermatic cord, typically occurs immediately after puberty. It is said to feel “like a bag of worms.” This disorder is caused by failure of the valves in the spermatic veins to prevent the backflow of blood. More than 95% of varicoceles occur on the left; the presence of a unilateral right-side varicocele indicates obstruction or compression of the inferior vena cava. A varicocele reduces testicular blood flow, which can lead to decreased spermatogenesis and infertility. Surgical ligation or inducement of slight sclerosis of the vein corrects the varicocele.

Spermatocele, a swollen aggregation of sperm in the epididymis, is difficult to distinguish from a hydrocele by palpation. Diagnosis is based on aspiration of fluid from the mass. The fluid in a hydrocele is clear, while the fluid in a spermatocele is milky and contains sperm cells. Spermatoceles are usually caused by inflammation and tend to resolve spontaneously. Most are asymptomatic, and unless discomfort or pain occurs, excision is not recommended.

Testicular Disorders

Disorders of the testis include ectopic testis, cryptorchidism, testicular torsion, and testicular cancer. Because the testes are the sites of spermatogenesis and androgen production, these disorders have potentially negative effects on both fertility and secondary sex characteristics.

Cryptorchidism and Ectopic Testis

Cryptorchidism and **ectopic testis** are similar congenital conditions in which descent of the testis is incomplete (in the first case) or has taken an abnormal route (in the second). In cryptorchidism, the descending testis stops in the pelvic cavity, inguinal canal, or upper end of the scrotum. An ectopic testis ends up in the perineal or suprapubic region or just beneath the skin of the thigh. Cryptorchidism can result from adhesions along the path of descent, an inguinal canal too narrow for the testis to pass through, an absent or abnormal gubernaculus, a too-short spermatic cord, and other causes. Ectopic testis results from a gubernaculum connected to the wrong site at its distal end, so that it pulls the descending testis to an abnormal location. Palpating the scrotum during the physical examination reveals the absence of the testis. Imaging techniques such as ultrasound, MRI, and CT scans are used to locate testes that cannot be found by palpation.

In both disorders, the testis is hormonally functional but is too warm for spermatogenesis. Fertility may be normal if only one testis is affected, but in bilateral cases the patient is sterile. If the condition is not corrected, the risk of testicular cancer is 35 to 50 times higher than normal. An ectopic testis is normally removed to prevent cancer. The surgical removal of a testis is called **orchiectomy.** In cryptorchidism, an injection of human chorionic gonadotropin can often stimulate completion of descent. If this fails, the testis is surgically relocated to the scrotum in young boys (a procedure called **orchiopexy**), but because of the cancer risk, orchiectomy is performed in boys over 10 years old and in men.

Testicular Torsion

Testicular torsion, twisting of the testis on the spermatic cord, may occur spontaneously or as a result of trauma or strenuous exercise. It occurs most often during puberty, but may happen at any time of life. Twisting the testicular blood vessels causes testicular ischemia, and if left untreated, necrosis of the testis. Testicular torsion also produces scrotal swelling that is not alleviated by rest or support of the testis. The patient complains of severe pain and nausea, and may vomit. Fever is common. Treatment is aimed at alleviating the torsion by manually rotating the affected testis or by surgery. Unless corrected within 4 to 6 hours, damage to the testis can permanently impair fertility.

Testicular Cancer

Testicular cancer is relatively rare, accounting for less than 1% of all cancers in men. It is also one of the most treatable cancers, with a cure rate exceeding 95%. Nevertheless, it causes about 350 deaths per year in the United States, mostly because it is not detected or treated in time to prevent metastasis. Up to 10% of men with testicular cancer are asymptomatic. Most victims are 15 to 34 years old. White men have about four times the incidence of testicular cancer that black men do, and within both groups, men of higher economic status have higher rates of testicular cancer. Some other risk factors for testicular cancer are high androgen concentrations and heredity. Brothers and especially identical twins show higher shared incidence of testicular cancer. The best defense against testicular cancer is testicular self-examination (TSE). This is best done after a shower because heat relaxes the scrotum. The testes should be gently rolled between the thumb and fingers, feeling for suspicious lumps. A slight inequality in size is normal. TSE should be as routine for men as breast self-examination (BSE) is for women.

Most testicular tumors arise from germ cells. The first sign of a tumor is an enlarged but often painless testis. Its continued growth causes lower abdominal aching or a feeling of testicular "heaviness." Epididymitis, gynecomastia, or hydrocele may develop. Because up to 25% of cases are misdiagnosed at first, testicular cancer often metastasizes, especially to the lungs, lymph nodes, and central nervous system. Metastasis to the lungs causes cough, dyspnea, and hemoptysis. Neural effects range from alterations in vision and mental status to seizures.

Diagnosis is achieved by palpation, imaging, and blood tests for such serum markers as α-fetoprotein, HCG, and lactate dehydrogenase. The affected testis is surgically removed, and metastasis is treated with radiation and chemotherapy. The prognosis depends on the tumor type and degree of metastasis. Most deaths occur within 2 years, and a disease-free survival time of 3 years is considered a cure. As evidence that men who survive testicular cancer can live a normal life, consider Lance Armstrong, one of the best professional cyclists in the world. After diagnosis and treatment for testicular cancer, Armstrong returned to training. In both 1999 and 2000, he won the Tour de France, a grueling 3-week, 2,300-mile cross-country bicycle race.

Prostatic Disorders

The most common disorders of the prostate are prostatitis, benign prostatic hyperplasia, and prostate cancer.

Prostatitis

Prostatitis, or inflammation of the prostate, is seen in up to 36% of males in the United States. It occurs in bacterial and nonbacterial forms. *Bacterial prostatitis* is usually caused by *Escherichia coli*, *Pseudomonas*, and *Streptococcus faecalis*. Its signs and symptoms are identical to those of a urinary tract infection, including dysuria, frequent urination, nocturia, and a weak urine stream. The patient may have a fever, fatigue, and pain in the joints, muscles, lower back, or rectum. The prostate is firm, swollen, tender, and painful. Treatment employs broad-spectrum antibiotics for up to 42 days, analgesics, bed rest, and ample water intake.

Nonbacterial prostatitis is more common and has an unknown cause. It, too, is characterized by symptoms similar to those of a urinary tract infection, along with pain in the infrapubic, suprapubic, scrotal, penile, or inguinal regions and pain upon ejaculation. Diagnosis is based on the absence of an infectious agent in urine cultures or prostatic fluid and inflammation of the prostate as confirmed by examination. Treatment varies, but usually includes bed rest, anti-inflammatory agents, and anticholinergics.

Benign Prostatic Hyperplasia

Benign prostatic hyperplasia (BPH), noncancerous enlargement of the prostate gland, occurs mostly in men over 50. Men of that age have a 25% to 30% chance of requiring a *prostatectomy* (removal of the prostate) at some time in their lives. The cause of BPH is still unknown, but the amounts of testosterone and other testicular steroids in the blood are thought to be contributing factors. The onset of BPH is slow. As the prostate increases in size, urethral compression causes the signs and symptoms of urinary tract obstruction—decreased force of the urine stream, increased frequency and urgency of urination, and nocturia. Over time, the obstruction worsens and the bladder cannot be fully emptied upon urination. This leads to incontinence or urine retention. Urine retention can cause elevated pressure in the kidneys *(hydronephrosis),* which presents a threat of renal failure.

BPH is diagnosed from these signs and symptoms, palpation, and laboratory analysis. Upon palpation, the prostate is found to be enlarged and to have lost its distinctive lobular shape. In up to 50% of patients, blood samples show an elevated level of prostate-specific antigen (PSA). Since the hyperplasia is not reversible, the only treatment in severe cases is surgical removal of the hyperplastic tissue or the entire prostate. In mild cases, treatment with androgen antagonists (such as Proscar) can reduce prostate size somewhat. Drugs that block α-adrenergic receptors and relax smooth muscle, such as prozosin and terazosin, have also been successful.

Prostate Cancer

Prostate cancer is responsible for almost 7% of all cancer deaths and is second only to lung cancer in cancer-related deaths in men in the United States. Although it accounts for 42% of all cancers in men, over 80% of all cases are seen in men aged 65 and over, and it rarely occurs before age 40. Diet and family history appear to influence the risk of developing prostate cancer. Androgens are not thought to cause prostate cancer, but they may promote tumor growth once the cancer has begun.

Most cases progress slowly and are asymptomatic at first, making early detection difficult. The initial signs and symptoms are those of urinary tract obstruction. Prostate cancer metastasizes to such sites as the pelvis, ribs, femur, vertebrae, lymph nodes, lungs, liver, and adrenal glands. Metastatic prostate cancer often produces bone pain and pathological fractures. Other signs include edema, hepatomegaly, and lymph node enlargement. Diagnosis is through DRE, blood screening for PSA, transrectal sonography, and biopsy. Sites of suspected metastasis are examined by MRI, CT, and biopsies.

Treatment depends on such factors as the stage of the cancer, anticipated side effects, and the patient's age, health, and life expectancy. Older men who are already in poor health may opt not to pursue treatment, and men concerned about side effects such as erectile dysfunction may also forgo treatment. Treatment includes hormonal, chemical, or radiation therapy, prostatectomy, or a combination. Side effects include incontinence and erectile dysfunction.

The average 5-year survival rate for treated prostate cancer is 78%; in cases without metastasis, it rises to 95%. Because early detection is a key to successful treatment, annual DRE and PSA screening are recommended for men over 50; only DRE is recommended for those between the ages of 40 and 50. However, men with a family history of prostate cancer are encouraged to start screening earlier, sometimes even in their 20s.

Case Study 27 The Athlete Who Ignored His Symptoms

Erik is a 21-year-old minor league hockey player hoping for a spot in the National Hockey League (NHL). In one game, Erik is involved in a fight and is struck in the groin by an opponent's stick. He falls to the ice in extreme pain and is taken to the hospital. In the emergency room, he is diagnosed with testicular torsion and a developing hydrocele. The physician relieves the torsion and aspirates a clear fluid from the hydrocele. Erik is told to visit his regular physician for further treatment if the swelling returns. He returns to play after a few days and finishes the season without incident. The following year, Erik is drafted by an NHL team.

Three years later, Erik develops difficulty breathing and a chronic cough that occasionally produces bloody sputum. Fearful that he could lose his place on the team, Erik downplays these symptoms when questioned by the team physician. But as the season progresses, the cough worsens and Erik begins to experience frequent headaches, dizziness, and blurred vision.

The team physician examines Erik and finds that he has lymphadenopathy in addition to his other symptoms. When questioned further, Erik mentions that he is becoming fatigued more easily and has been experiencing some abdominal pain.

During the physical examination, the team physician notes that Erik has a mass in the right testicle, but none in the scrotum. He asks Erik if the testicle had been giving him any trouble. Erik says that he's noticed it seemed somewhat enlarged, but he thought this was just because of the injury he received back in the minor leagues. Suspecting a testicular tumor, the doctor orders blood tests and a CT scan. The blood analysis shows elevated concentrations of lactate dehydrogenase and α-fetoprotein. The CT scan reveals seven lung tumors and a brain tumor. Erik is referred to an oncologist and diagnosed with metastatic testicular cancer. Erik's oncologist tells him that in cases this advanced, the chance of survival is about 50%. Erik is determined to beat the cancer, however, and agrees to an aggressive course of treatment.

He then undergoes orchiectomy and brain surgery to remove the primary testicular tumor and the metastatic brain tumor. He undergoes a successful year-long course of chemotherapy for the lung tumors. After remaining cancer-free for 2 years, Erik is told that if he remains so for one more year, he will be considered cured.

Based on this case study and other information in this chapter, answer the following questions.

1. If Erik's testicular torsion were to go untreated, what would happen to the testis?
2. Why is the swelling in his testis diagnosed as a hydrocele rather than a spermatocele or a varicocele?
3. Why does Erik's cancer go undiagnosed long enough to metastasize so widely?
4. Considering his occupation, what might happen to Erik if the cancer metastasizes to his bones?
5. Why do you think Erik's brain tumor is treated surgically, instead of being treated with chemotherapy like the lung tumors?
6. Angelo, a 55-year-old male, is diagnosed with cirrhosis of the liver. Which of the following disorders might he develop as a secondary effect?
   a. prostatitis
   b. testicular cancer
   c. spermatocele
   d. hydrocele
   e. gynecomastia

7. Terry is diagnosed at the age of 14 with an ectopic testis located in the groin. The family physician recommends orchiectomy rather than surgically relocating the testis to the scrotum. His parents, who would like to be grandparents one day, protest and ask if it wouldn't be better just to relocate the testis. How would you expect the doctor to advise them?
8. Why should men begin regular testicular self-examination (TSE) at an earlier age than they start receiving digital rectal examinations (DREs)?
9. Predict what a man might feel if he had epididymitis and did a TSE.
10. In what way does varicocele resemble varicose veins? (See chapter 20 of this manual.)

Selected Clinical Terms

benign prostatic hyperplasia (BPH) Noncancerous growth of the prostate gland, leading to urethral compression and impeded urine flow.

cryptorchidism The failure of one or both testes to completely descend through the inguinal canal into the scrotum.

digital rectal examination (DRE) Palpation of the prostate and neighboring structures by means of a gloved finger inserted into the rectum.

ectopic testis The descent of a testis to an abnormal location such as the perineal, suprapubic, or femoral region.

gynecomastia Abnormal enlargement of the male breasts.

hydrocele An accumulation of serous fluid in the tunica vaginalis of the scrotum.

orchiectomy Surgical removal of a testis.

orchiopexy Surgical translocation of a testis to its normal location in the scrotum.

phimosis The presence of a tight, nonretractable foreskin; a risk factor for penile cancer.

prostate-specific antigen (PSA) An antigen secreted by the prostate gland into the semen; elevated levels in the blood serum indicate prostatic enlargement and may be an early warning of prostate cancer.

prostatitis Inflammation of the prostate gland.

spermatocele A sperm-containing cyst in the epididymis.

testicular torsion Twisting of a testis on the spermatic cord, causing severe pain and testicular ischemia.

varicocele Abnormal dilation of veins in the spermatic cord due to failure of the venous valves to prevent backflow of blood.

28 The Female Reproductive System

Objectives

In this chapter we will study

- common symptoms of female reproductive disorders;
- diagnostic procedures for examining the female reproductive system;
- two menstrual disorders—dysmenorrhea and amenorrhea;
- ovarian cysts and polycystic ovary;
- inflammatory diseases of the female reproductive system;
- endometriosis; and
- cervical, endometrial, and ovarian cancer.

Assessment of the Female Reproductive System

Up to 20% of women who visit a physician have gynecological or obstetrical complaints. Menstrual irregularity is often the chief complaint—abnormally heavy or light flow, failure to menstruate, bleeding between periods, or painful menstruation. Such irregularities may be due to inflammation, sexually transmitted diseases (STDs), improper nutrition, or oral contraceptives. Pelvic pain and dysuria, symptoms seen in male reproductive disorders, routinely occur in females as well. Pelvic pain may be due to inflammatory disease, an ovarian cyst, or a tubal pregnancy. Dysuria and abnormal vaginal discharge are most often the result of a microbial infection. Women also seek medical attention for infertility.

As with male reproductive disorders, a thorough patient history, confidence in the physician, and strict confidentiality regarding personal information are vital to an accurate diagnosis and appropriate treatment and counsel.

Physical Examination

After a routine check of height, weight, and vital signs, a gynecological examination focuses on the breasts, abdomen, and pelvic area. The examiner palpates the breasts, reminds the patient to conduct monthly breast self-examination (BSE), and may recommend a mammogram. He or she also palpates the abdomen.

In the pelvic examination, the external genitalia are visually inspected for lesions, trauma, abnormal masses, or swellings. Abnormal discharges from the urethra or vagina are also noted, and samples are taken for culture and identification of pathogens. The walls of the vagina and the cervix are inspected by using a **speculum** to spread the vagina and make the mucosa more visible. Redness of the vaginal walls indicates inflammation *(vaginitis),* a common sign of infection. In nonpregnant patients, cyanosis of the vaginal walls often indicates pelvic tumors. (Cyanosis of the vaginal wall and cervix, called *Chadwick's sign,* is normal in pregnancy.) The cervix is checked for lacerations, ulcerations, polyps, and abnormal discharge. A small brush or spatula is used to collect cells from the external os and prepare a *Pap smear* for histological evaluation. Next, the clinician conducts bimanual palpation by inserting two gloved fingers into the vagina and using the other hand to palpate the uterus and ovaries through the abdominal wall. This yields information about the size, shape, location, and tone of the uterus, the enlargement or absence of an ovary, and the presence of any abnormal pelvic masses.

The final component of the examination is a **rectovaginal examination,** which is done by placing the index finger into the vagina and the middle finger into the rectum, and palpating the uterosacral ligaments and the back of the uterus and cervix. The examiner checks for abnormal masses and the proper placement and tone of the ligaments. If a woman is pregnant, care is taken to minimize disrupting the placenta or causing fetal trauma.

Laboratory and Imaging Tests

When warranted by the findings of the physical examination, samples of blood, urine, and vaginal and urethral discharge can be evaluated in the laboratory. Blood is analyzed for levels of reproductive hormones such as estradiol, progesterone, luteinizing hormone, follicle-stimulating hormone, and prolactin. Urine is assayed for estradiol and progesterone metabolites. If pregnancy is suspected, the blood or urine sample can be analyzed for the presence of HCG, which is detectable only in pregnant women. Samples of blood and vaginal or urethral discharge can be analyzed for the presence of pathogens. The female reproductive tract can also be examined by X ray, laparoscopy, CT, MRI, and sonography. Mammography (X-ray examination of the breasts) should be done biannually in the 40s and annually after age 50.

Menstrual Alterations

Menstrual disorders include painful menstruation *(dysmenorrhea),* lack of menstruation *(amenorrhea),* and *dysfunctional uterine bleeding,* which means abnormal duration, frequency, or timing of menstruation. Only dysmenorrhea and amenorrhea are discussed here.

Dysmenorrhea

Dysmenorrhea affects half to three-quarters of girls and women from 15 to 25 years of age but declines after the mid-20s. It often causes women to miss work or school. *Primary dysmenorrhea* is painful menstruation that cannot be traced to any underlying pelvic disease. *Secondary dysmenorrhea* is traceable to some other reproductive system pathology. In both forms, the endometrium produces excessive amounts of prostaglandin, which in turn causes constriction of the uterine blood vessels, endometrial ischemia, and painful contractions of the uterine muscle. The pain begins in the pelvic region, radiates to the groin, and is often accompanied by backache, anorexia, vomiting, diarrhea, headache, and syncope. Dysmenorrhea does not occur when a woman does not ovulate, and is therefore relieved by oral contraceptives. Exercise, a low-fat diet, heat and massage, relaxation techniques, and orgasm also relieve the symptoms.

Amenorrhea

Amenorrhea has a variety of causes and can be broadly categorized according to the patient's menstrual history. *Primary amenorrhea* is a failure of the menses to begin. A girl is considered amenorrheic if she reaches the age of 14 without menarche or the development of secondary sex characteristics, or if she reaches 16 without menarche even if she does develop the secondary sex characteristics. There are numerous causes of primary amenorrhea, including genetic disorders such as Turner syndrome, congenital CNS defects such as hydrocephalus, anatomical abnormalities such as lack of a uterus or vagina, and CNS lesions. *Secondary amenorrhea* is the cessation of menses in women who have previously menstruated. It is considered to exist when a women misses three consecutive periods. It can result not only from a variety of diseases but also from extreme weight loss, either through malnutrition (as in people with anorexia nervosa) or through exercise (as in dancers, gymnasts, and runners). Secondary amenorrhea is also normal during pregnancy, lactation, and menopause, and is common in early adolescence.

Amenorrhea is diagnosed primarily from the patient history, but the underlying cause may require laboratory analyses such as blood hormone assays or imaging methods to look for tumors. Menstruation often resumes spontaneously (for example, when breast-feeding is terminated), or it may require weight gain or treatment of an underlying disease. Menstruation can sometimes be restored by hormone replacement therapy (HRT).

Ovarian Cysts and Polycystic Ovary

Ovarian cysts usually occur in women between the ages of puberty and menopause, but may appear before or after those stages. They are benign cysts of either follicular or luteal origin, arising from an ovarian follicle that fails to ovulate or degenerate or from an abnormally persistent corpus luteum. Both types are hormonally active—follicular cysts produce estrogen, and luteal cysts produce progesterone. Ovarian cysts typically grow to 5 or 6 cm in diameter, but sometimes up to 10 cm. They produce abdominal and back pain, **dyspareunia** (pain during intercourse), and irregular menstruation. Luteal cysts are less common but cause more symptoms, including pelvic pain, amenorrhea, or delayed but heavy menstruation. They occasionally rupture, causing excruciating pain and requiring emergency surgery. Most ovarian cysts, however, require no specific treatment and regress spontaneously within 2 months.

Polycystic ovary (PCO) affects women mainly between the ages of 15 and 30. It results from multiple follicles that develop but never ovulate, so the ovaries accumulate numerous cysts about 2 to 6 mm in diameter. The underlying cause is a positive feedback cycle of abnormal hormone secretion. FSH level is low and LH level is high. The high LH level causes the adrenal cortex and ovary to secrete an abnormally high level of androgens, which various tissues convert to estrogen. The high estrogen level inhibits FSH secretion and stimulates LH secretion, thus completing the positive feedback loop. There is enough FSH to cause more and more follicles to begin developing, but not enough to cause them to mature and ovulate. Thus, they turn into ovarian cysts.

Women with PCO typically present with obesity, excessive growth of body and facial hair (hirsutism), and amenorrhea, although others lack hirsutism and obesity and complain instead of irregular, heavy uterine bleeding, an effect of the high estrogen level. PCO is diagnosed from the patient history, physical exam, and high levels of androgen and estrogen in the blood. The treatment for PCO depends on whether the woman wants to become pregnant and other circumstances of individual cases. Options include antiestrogenic drugs, progesterone, and low-dose oral contraceptives. For hirsutism, bleaching of the hair or hair-removal methods such as waxing or electrolysis are recommended. If all else fails, ovarian surgery may be used to remove some or all of the cysts.

Inflammation of the Female Reproductive System

The female reproductive tract is subject to numerous inflammatory disorders, caused by a variety of infectious agents and irritants. These disorders include inflammation of the vulva *(vulvitis),* vagina *(vaginitis),* cervix *(cervicitis),* uterine tubes *(salpingitis),* ovaries *(oophoritis),* or a combination of the uterus, uterine tubes, and sometimes the ovaries *(pelvic inflammatory disease).* In many of these diseases, any sexual partners should be treated as well.

Vulvovaginitis is a combination of *vulvitis* and *vaginitis,* often stemming from irritation by soap, lotion, menstrual pads or tampons, feminine hygiene sprays, or tight clothing. However, the main causes are STDs and infection by *Candida albicans* ("yeast infection"). The incidence of sexually transmitted vulvovaginitis is greatest in women between the ages of 10 and 24. Vulvovaginitis produces dysuria, redness of the vulva, and a purulent, foul-smelling vaginal discharge. The discharge can be cultured to confirm the diagnosis and determine what antibiotic, if any, should be used in treatment. Medications for vulvovaginitis also include agents to suppress inflammation and acidify the vaginal pH. Patients are counseled about hygiene and causative irritants. If the condition is caused by an STD, the woman's sexual partner also must be treated.

Cervicitis usually results from an STD. The cervix becomes red and edematous, and affected women complain of dysuria, pelvic pain, and bleeding. Some also report a mucopurulent (mucus- and pus-containing) discharge. Diagnosis is based on these symptoms and identification of the causative agent in the discharge. Treatment employs antibiotics specific for the pathogen.

Pelvic inflammatory disease (PID) usually stems from STDs in which the microbes have migrated from the vagina to the uterus, uterine tubes, and ovaries. Most cases of PID occur in sexually active women under the age of 35 and are caused by a mixed microbial infection. The scarring induced by PID raises the risk of infertility or ectopic pregnancy. Death is rare (8% to 9% of cases), but when it occurs it is usually due to septic shock. Women may be asymptomatic or show a wide range of signs and symptoms, including extreme abdominal pain, guarding or rebound tenderness (see chapter 25 of this manual), and discomfort upon movement of the cervix. Fever, vomiting, leukocytosis, and large quantities of a purulent cervical discharge are also present. Diagnosis is based on the patient history, signs and symptoms, and identification of the causative agent. Treatment includes bed rest, antibiotics, and sexual abstinence during treatment.

Endometriosis

Endometriosis is the ectopic growth of endometrial tissue. That is, endometrial tissue spreads beyond the uterus and grows in the uterine tubes or on the ovaries, uterine ligaments, and other sites. The ectopic tissue is responsive to the reproductive hormones and undergoes the same cyclic changes that would be seen in the uterus (proliferation, secretion, and bleeding). Endometriosis is thought to occur in 3% to 10% of women of reproductive age, but many of them are asymptomatic and the endometriosis is discovered in the course of examining some other condition. There is a genetic predisposition to endometriosis, but otherwise the cause is unknown.

The signs and symptoms include pelvic pain, dysmenorrhea, constipation, dyspareunia, *dyschezia* (painful defecation), abnormal vaginal bleeding, and infertility. Diagnosis is based on the symptoms, family history, laparoscopy, and biopsy. Treatment is aimed at alleviating pain, halting the progress of the disease, and restoring fertility. Treatment strategies depend on the patient's age and desire to have children in the future. Options include hormones such as oral contraceptives to suppress growth of the ectopic tissue, surgical removal of as much of the ectopic tissue as possible, and surgical removal of the uterus and ovaries followed by hormone replacement therapy. Unfortunately, nearly half of the women treated experience a recurrence of endometriosis within 5 years.

Cancers of the Female Reproductive System

Breast cancer strikes 1 out of every 8 or 9 women in the United States and is a leading cause of female mortality. Other cancers of the reproductive system account for 1 in 8 cases of cancer and 1 in 10 cancer-related deaths in women of the United States. Here we consider three of these: cervical, endometrial, and ovarian cancer.

Cervical Cancer

Cervical cancer is the second most common cancer of women worldwide and the fourth most common in the U.S. It is now considered a sexually transmitted disease because 75% to 90% of cases result from infection with the human papillomavirus (HPV), which is transmitted by intercourse. Risk factors for cervical cancer include having first intercourse before the age of 16, having multiple sexual partners or a male partner who has had multiple partners, and smoking.

The stages of cervical pathology are classified according to histological changes. It begins as *cervical intraepithelial neoplasia (CIN*; also known as *cervical dysplasia*). The dysplasia is considered mild, moderate, or severe according to the depth to which it extends in the epithelium. CIN progresses to cervical cancer in 10% to 75% of cases, taking 10 to 12 years to become malignant. CIN is asymptomatic but can be recognized in a Pap smear, which underscores the importance of regular Pap smears and early detection.

Cervical carcinoma (cancer) produces vaginal bleeding; a clear, pink, or yellowish vaginal discharge; an abnormal, sometimes foul vaginal odor; and abnormal menses. Diagnosis can be confirmed by a Pap smear. Treatment depends on the stage and size of the tumor. Precancerous lesions are often ablated by **cryosurgery** or **laser surgery.** If the lesion is cancerous, the treatment modalities are radiation, chemotherapy, and hysterectomy. With early detection, precancerous lesions have a cure rate approaching 100%, while the overall 5-year survival rate is 67%.

Endometrial Cancer

Endometrial cancer accounts for 13% of all cancers in women. It usually occurs after menopause, with peak incidence between the ages of 58 and 60. At greatest risk are women who are obese, anovulatory, have a family history of breast or ovarian cancer, or had an early menarche or late menopause. The most common symptom is abnormal vaginal bleeding. In fact, more than 30% of all postmenopausal women with vaginal bleeding have endometrial cancer. Later in the disease, pain and weight loss are also seen. Screening and diagnosis are done through endometrial biopsy. Additional imaging techniques, such as CT, MRI, bone scans, ultrasound, and barium enema, are used to locate the metastatic sites.

As with cervical cancer, early identification is the key to achieving a cure, and treatment depends on the stage of the disease. All tumors are removed through either **curettage** or **hysterectomy.** Chemotherapy and radiation therapy are also used. The prognosis depends on the severity of the disease and the age and health of the patient. Five-year survival rates approach 65% for all cases.

Ovarian Cancer

Ovarian cancer has the highest incidence in industrialized countries, accounting for 18% of all reproductive cancers. Half of all women with ovarian cancer are over 65 years of age, but it can strike at any age. The causes of ovarian cancer are currently unknown, but it is most common in women who have never had children, who have had breast cancer, or whose first-degree relatives (a mother, daughter, or sister) have had ovarian cancer. Ovarian cancer can arise from either the surface epithelium or the germ cells of the ovary. Epithelial cancers affect mainly women over 50. Germ cell tumors, on the other hand, arise from the primitive germ cells of the fetal ovary and occur mostly in children and adolescents.

The signs and symptoms vary considerably. Unfortunately for most women, by the time noticeable signs or symptoms appear, metastasis has already occurred. The first symptoms are usually abdominal pain and swelling. The patient may experience dyspnea, vomiting, and inappropriate vaginal bleeding. Unlike cervical and endometrial cancer, screening tests for ovarian cancer do not exist. Diagnosis is based on biopsy and imaging techniques such as ultrasound, CT, and MRI. Treatment involves surgery followed by chemotherapy and radiation therapy. The 5-year survival rate for all stages of ovarian cancer is 42%. If detected early, survival rates exceed 90%, but when metastasis has occurred, they decrease to between 20% and 40%.

Case Study 28 The Woman with Abdominal Pain

Cathy, a 20-year-old college sophomore, visits the student health center complaining of nausea, vomiting, and extreme abdominal pain. She says the pain came on suddenly after a meal, so she is concerned about food poisoning. During the patient history, Cathy discloses that she is sexually active and has more than 10 sexual partners per year. She uses oral contraceptives as her primary method of birth control and does not rely on her partners to use condoms. Cathy has a past history of vulvovaginitis, cervicitis, and numerous STDs (chlamydia, syphilis, and genital herpes). Recently she has noticed a whitish vaginal discharge with a strong odor that has increased in quantity over the past week. Cathy's vital signs are as follows:

Oral temperature = 101.4°F (38.6°C)

Heart rate = 78 beats/min

Respiratory rate = 15 breaths/min

Blood pressure = 150/86 mmHg

A gynecological examination is performed. Palpation of the abdomen reveals abdominal guarding, rebound tenderness, and an enlarged, painful uterus. A surgical scar is also noted, and Cathy explains that her appendix was removed when she was a child. The external genitalia appear slightly edematous but are otherwise normal. Both the vagina and the cervix are slightly inflamed, and a purulent discharge is noted. A sample of the discharge is taken for culture, and a cell sample is taken from the external os for a Pap smear. During the pelvic examination, movement of the cervix creates abdominal discomfort, and the rectovaginal examination confirms the uterine enlargement noted upon palpation.

Blood, fecal, and urethral samples are obtained. Blood tests reveal leukocytosis but no HIV antibodies. The Pap smear is negative, and the cultures are positive for *Chlamydia* and *Neisseria gonorrhoeae*. A pregnancy test (HCG assay) is negative. Ultrasound imaging of the pelvic cavity shows enlargement of the uterus and uterine tubes without pregnancy. Based on these test results, Cathy is diagnosed with pelvic inflammatory disease.

Cathy is prescribed a combination of antibiotics and bed rest for 10 days. During this treatment period, she is told to refrain from intercourse. She is also advised to notify her partners about her condition and to encourage them to seek treatment. In addition, Cathy is told to return for a follow-up examination after completing her medication to ensure that the infections have been controlled. Finally, Cathy is encouraged to insist on condom use to minimize her chances of contracting STDs in the future. Because of her history of numerous STDs, she is warned that she is at increased risk for infertility due to uterine tube scarring as well as for uterine and cervical cancer.

Based on this case study and other information in this chapter, answer the following questions.

1. Which of Cathy's signs and symptoms are common to both pelvic inflammatory disease and appendicitis? Why is it important to rule out appendicitis?
2. What signs and symptoms support the diagnosis of PID?
3. Suppose Cathy says that in spite of her "wildness" in college, she just wants to "get it all out of my system, and then settle down and have kids after I graduate." As her physician, what would you tell her about the relevance of her present behavior to her family plans?
4. Why are a Pap smear and an HIV test conducted?
5. Why do Cathy's signs and symptoms rule out cervical, endometrial, and ovarian cancer?

6. Erica, a 17-year-old female, is a member of her high school swim team. She visits her physician because she has missed her last two menstrual periods. Laboratory tests are negative for HCG. Based on this limited information, speculate about the cause of Erica's amenorrhea. What treatment, if any, would you recommend?

7. Twenty-eight-year-old Brenda visits her gynecologist complaining of irregular menstruation, back pain, and pain during intercourse. Which of the following signs would suggest that Brenda is suffering from polycystic ovary disease?
   a. elevated serum LH concentrations
   b. elevated serum HCG concentration
   c. mucopurulent discharge
   d. rebound tenderness
   e. dyschezia

8. Andrea, a 32-year-old female, visits her nurse practitioner complaining of constipation, pelvic pain, and dysmenorrhea. Within the last 6 months, Andrea's sister has been diagnosed with endometriosis. Laparoscopy reveals two regions of ectopic endometrial tissue on Andrea's ovaries. Since Andrea does not intend to have any children, she decides to take a GnRH agonist in an attempt to treat the disorder without resorting to surgery. Explain how this medication would help alleviate the signs and symptoms of this disorder.

9. Why is maintaining an acidic pH in the vagina important in treating vaginitis?

10. A 15-year-old girl misses 3 days of school due to severe menstrual cramps before her mother takes her to a gynecologist. If you were the gynecologist, what issues would you discuss with the girl while taking her history? Suppose the gynecologist tells the girl simply to use a heating pad and analgesics, wait it out for another day or two, and return to school when she feels better. What do you think the diagnosis is?

Selected Clinical Terms

amenorrhea Absence of menstruation; classified as *primary* if an adolescent girl never begins menstruating and *secondary* if a previously menstruating woman ceases to menstruate for three or more consecutive cycles.

cervicitis Inflammation of the cervix of the uterus.

cryosurgery An operation to destroy tissue by freezing it with liquid nitrogen or liquid carbon dioxide.

curettage Scraping the internal lining of a hollow organ, such as the uterus, to remove new growths or to obtain tissue for biopsy.

dysmenorrhea Difficult or painful menstruation.

dyspareunia Pain experienced during sexual intercourse.

endometriosis The growth of endometrial tissue in ectopic sites such as the pelvic cavity or uterine tubes.

hysterectomy Surgical removal of the uterus.

laser surgery An operation to destroy unwanted tissue or cauterize an organ by burning it with a laser beam.

ovarian cyst A cyst, often several centimeters in diameter, that develops from an ovarian follicle or corpus luteum.

pelvic examination Physical examination of the external genitalia, vagina, cervix, and internal reproductive organs by visual observation and palpation, for purposes of health assessment or diagnosis.

pelvic inflammatory disease (PID) Inflammation in the pelvic cavity, usually caused by sexually transmitted microbes that migrate up the uterus and uterine tubes into the cavity.

polycystic ovary (PCO) The accumulation of numerous ovarian cysts arising from follicles that partially develop but fail to ovulate.

speculum Any instrument designed to be inserted into a body cavity or canal and used to spread it for easier viewing; varieties exist for dilating the vagina, rectum, eyelids, nostrils, and auditory canals.

vulvovaginitis Inflammation of the vulva and vagina.

29 Human Development

Objectives

In this chapter we will study

- the common signs and symptoms of pregnancy;
- some complications of pregnancy;
- the effects of drug use during pregnancy;
- prenatal testing for congenital anomalies;
- the elements of neonatal assessment;
- the causes and characteristics of precocious and delayed puberty; and
- how life expectancies and causes of death have changed in modern developed nations.

Assessment During Pregnancy

The estimated date of conception, anticipated date of birth, and progress of a pregnancy are determined from anatomical and physiological changes that occur in a woman's body during gestation. Typically, the earliest sign of pregnancy is amenorrhea—missing a menstrual period or two. In addition, a woman in early pregnancy may experience fatigue, breast enlargement, and nausea. Pregnancy is typically confirmed by means of chemical assays for human chorionic gonadotropin (HCG), a hormone that begins to appear in the blood and urine about 10 days after conception. Over-the-counter pregnancy testing kits and pregnancy tests administered in the clinic are both based on an HCG assay. HCG concentration rises rapidly in early pregnancy, doubling in concentration every 48 hours for the first 60 days.

Three signs have traditionally been accepted as proof of pregnancy: (1) fetal heart sounds; (2) fetal movements; and (3) identification of a fetal skeleton. More recently, however, proof of pregnancy has been expanded to include the doubling of HCG concentration and sonographic detection of the amniotic sac and fetus. Palpation of the fundus (superior curvature) of the uterus helps establish the gestational timetable. By 12 weeks of gestation, the uterus exceeds the volume of the pelvic cavity and extends into the abdomen. At 20 weeks, the fundus is at the level of the woman's umbilicus, and by 36 weeks, it can be palpated near the xiphoid process.

Complications of Pregnancy

A number of complications can occur during pregnancy and jeopardize the life and development of the fetus. Some common complications are spontaneous abortion, ectopic pregnancy, abruptio placentae, placenta previa, preeclampsia, and eclampsia.

Spontaneous abortion *(miscarriage)* has multiple definitions. In this manual, we use the term to mean delivery of the fetus before the twentieth week of pregnancy, while we consider delivery between 20 and 38 weeks *preterm birth.* Up to 15% of pregnancies end in spontaneous abortion, and more than half of these are associated with gross fetal abnormalities. In such cases, spontaneous abortion may be an adaptation of the maternal body to detect fetal abnormalities and minimize wasted investment in producing a child that would not live long after birth. About 85% of spontaneous abortions occur in the first 3 months of gestation. Spontaneous abortions begin with cramping and bleeding; in fact, up to 30% of women experience cramping or bleeding during the first 20 weeks of pregnancy. Some of these cases resolve without treatment; others require treatment to prevent abortion; and still others end in spontaneous abortion despite treatment. Treatment is bed rest and, in some cases, medications to minimize uterine contractions. The uterus and conceptus are monitored by sonography. If fetal cardiac activity is absent, the contents of the uterus are evacuated.

Ectopic pregnancy results when a conceptus implants anywhere other than in the uterus. Implantation can occur almost anywhere there is adequate blood supply to support fetal development, such as the cervix or the external surface of the uterus, colon, or ovary. However, the most common site is a uterine tube *(tubal pregnancy).* Untreated tubal pregnancy is usually fatal, because the uterine tube cannot expand enough to accommodate the growing conceptus. It ruptures within 12 weeks (usually around 6 to 8 weeks), causing sudden, intense lower abdominal pain, fainting, and sometimes death from hemorrhagic shock. One out of every 300 pregnancies is ectopic; the condition is fatal to the mother in 1 out of 826 cases in the United States. The first symptoms are cramping and spotting, usually after the first missed menstruation. Ectopic pregnancy is suspected on the basis of these signs and symptoms, and can be detected by palpation and confirmed by sonography. It requires surgical removal of the conceptus, with an effort to save and reconstruct the tube if possible.

Abruptio placentae is the premature detachment of the placenta from the uterine wall. It occurs in as many as 3.5% of pregnancies, usually in the last trimester. Risk factors include hypertension, other cardiovascular diseases, and rheumatoid diseases; it is also frequently induced by cocaine use. Vaginal bleeding is a common sign, but in some cases the blood is retained internally *(concealed hemorrhage).* Abruptio placentae is also suggested by a tight, tender uterus, maternal shock, or fetal cardiac distress, and it can be confirmed by sonography. Bed rest is sufficient treatment for mild cases, but in severe or worsening cases, the infant must be delivered, usually by cesarian section.

Placenta previa is attachment of the placenta in a low position in the uterus, near or obstructing the internal os of the cervix. It occurs in about 1 out of 200 pregnancies. The difficulty presented by placenta previa is that the infant cannot be delivered vaginally without the placenta separating from the uterine wall first, thus cutting off the fetal blood supply and jeopardizing the life of the fetus. Placenta previa is difficult to distinguish from abruptio placentae and is treated similarly—by bed rest if mild and cesarian delivery if more severe.

Preeclampsia is a serious pathology, largely cardiovascular in nature, that occurs in about 4% of pregnancies. It occurs especially in the third trimester and affects mainly *primigravidas* (women pregnant for the first time) and women with a history of hypertension or other vascular disease. The major complications of preeclampsia are abruptio placentae and the likelihood of developing into a more severe, potentially fatal condition, *eclampsia,* described next. Mild preeclampsia can be treated on an outpatient basis, but the patient must see her physician every 2 days and is normally admitted to a hospital if the condition does not quickly improve. Since the etiology of preeclampsia is unknown, treatment is aimed at lessening the symptoms and preserving the life of the mother. The fetus usually survives if the mother does. Bed rest and I.V. saline are used, and magnesium sulfate ($MgSO_4$) or other anticonvulsive agents are given to help lower the blood pressure and prevent convulsions (see eclampsia). If the fetus is of viable age, it is delivered as soon as practical.

Eclampsia typically develops from untreated preeclampsia. It is distinguished by convulsions, coma, or both and is often fatal to the mother, fetus, or both. A woman with eclampsia must be monitored very closely. $MgSO_4$ is administered, and the baby is delivered as soon as possible, by cesarian section if necessary. Eclampsia often occurs even in the postpartum period, usually within 4 days. Patients who were preeclamptic are monitored closely for 6 to 8 weeks, with special attention to blood pressure, CBC, BUN, urinalysis, and creatine level.

Drug Use During Pregnancy

As of 1997, up to 90% women surveyed by the Centers for Disease Control admitted to using over-the-counter prescription medications or other drugs during pregnancy. These included antacids, analgesics, antihistamines, antiemetics, tranquilizers, hypnotics, socially accepted drugs (alcohol, nicotine, caffeine), and drugs of abuse (cocaine, marijuana, LSD). What many of these women fail to realize (or decide to ignore) is that drugs in their circulation have access to the fetal circulation. During pregnancy, they can affect the fetus in a number of ways—for example, by producing a toxic or teratogenic effect on the embryo or fetus and by altering placental function, myometrial activity, and maternal metabolism. The degree to which a given drug affects the fetus depends on the amount (dose) and potency of the drug and the age of the conceptus. The conceptus is especially vulnerable to teratogenic effects between 3 weeks and 8 weeks of gestation, but even after the eighth week, drugs influence the growth and function of the fetal organ systems.

The United States Food and Drug Administration has developed a universally accepted labeling system describing the effects of drugs on the fetus. Drugs are placed in one of five categories:

- Category A drugs present no fetal risk.
- Category B drugs have been shown to affect fetal development in other animals but not in humans.
- Category C encompasses drugs for which adequate studies have not been conducted and drugs whose adverse effects on animals have not yet been studied in humans.
- Category D drugs are known to affect human fetal development, but have benefits that sometimes outweigh this risk.
- Category X drugs have well-established deleterious effects on the human fetus that outweigh any possible benefits.

Of the socially accepted drugs, the one with the most effect on fetal development is alcohol. *Fetal alcohol syndrome (FAS)* affects babies whose mothers drink excessive amounts of alcohol (more than three drinks per day) during pregnancy. The mother's consumption of alcohol during pregnancy can have effects that last throughout her child's life. The fetus is often born with such defects as microcephaly, joint anomalies, cardiovascular defects, and shortened palpebral fissures. Impaired CNS development leads to mental retardation, while alterations in digestive system function result in decreased growth and failure to thrive.

People who drink caffeinated beverages may laugh at the thought of caffeine being a "drug." However, an increasing amount of evidence suggests that high caffeine intake (the equivalent of 7 to 8 cups of coffee per day) has deleterious effects on a fetus. While these studies are still preliminary, they suggest that caffeine may increase the incidence of miscarriage, preterm birth, spontaneous abortion, and low birth weight.

The exact carcinogenic or teratogenic effects of cigarette smoke are not known. What is known is that smoking decreases the partial pressure of oxygen in maternal blood and thus reduces the amount of oxygen available to the fetus. Infants born to smoking mothers weigh as much as 6 ounces less than infants born to nonsmokers and may exhibit slower than average postpartum development. Smoking has also been implicated in an increased risk of spontaneous abortion, stillbirth, preterm birth, and infant mortality.

Testing for Congenital Anomalies

Congenital anomalies (birth defects) result from alterations in fetal development. A number of them have been discussed in previous chapters. Some abnormalities have been linked to environmental factors such as air and water pollution, the use of drugs during pregnancy, and spontaneous genetic mutations.

Parents with a family history of a particular genetic disease are advised to consult a genetic counselor to determine their risk of conceiving a child with the disorder. Placental sampling and other prenatal screening techniques allow some genetic diseases to be identified after conception. For example, an elevated concentration of α-fetoprotein in the maternal circulation is associated with neural tube defects and other congenital anomalies. Fetal cells can be obtained for genetic analysis by skin sampling, umbilical blood sampling, amniocentesis, or chorionic villus sampling. Finally, sonography can identify a number of gross congenital abnormalities. Parents can then use this information to either prepare for a child with special needs or consider a therapeutic abortion.

Neonatal Assessment and Care

Assessment of the health of the newborn (neonate) is carried out in slightly different ways in different health-care facilities. In general, the neonate is first inspected in the delivery room to identify any life-threatening conditions or gross abnormalities. The *Apgar score* is determined at 1 and 5 minutes after birth; the heart and lungs are auscultated, abdominal organs palpated, and sex established. The baby is weighed and measured, and its gestational age is estimated from the weight, length, and head circumference. The baby is wrapped in a blanket, and 1% silver nitrate or antibiotic solution is placed in each eye as a precaution against congenital infections. The mother is allowed to hold (and nurse) the neonate if she wishes before it is taken to the nursery.

In the nursery, the infant is bathed after its temperature stabilizes, and it is given a shot of vitamin K (phylloquinone) to promote a healthy prothrombin level and efficient blood clotting. (The relationship between vitamin K and blood clotting was discovered in Denmark, where *coagulation* is spelled with a *k* instead of a *c;* hence the name of the vitamin.) Skin color and vital signs are noted. The head and genitalia are inspected for abnormalities, the abdominal organs palpated, and reflexes tested. Blood tests are conducted to check the bilirubin level and test for PKU, sickle-cell disease, anemia, and other disorders.

Precocious and Delayed Puberty

Puberty is normally evident by the ages of 8 to 13 in girls and 9 to 14 in boys. Its earliest signs are thelarche (breast development) and testicular enlargement. Appearance of these signs earlier than these ranges is called *precocious puberty,* and if later, *delayed puberty.*

Precocious puberty is more common in girls than in boys. Puberty begins as young as age 3 in a significant number of girls. History's youngest mother, a Peruvian girl named Lina Medina, began menstruating at the age of 8 months; her breasts and pubic hair developed by the age of 4 years; and she gave birth to a 6-pound boy when she was 5 years and 7 months old. Precocious puberty is usually idiopathic but involves either premature secretion of pituitary gonadotropins or secretion of the gonadal sex steroids in the absence of gonadotropin stimulation. A child exhibiting precocious puberty is thoroughly examined for life-threatening tumors of the CNS, gonads, or adrenal glands. Precocious puberty is managed by removing the underlying cause, if known, or by hormone therapy to reverse sexual development or retard it until an appropriate age. Contraception and emotional counseling are also important elements of patient care.

Delayed puberty is more common in boys than in girls and is often hereditary. In most cases, the hypothalamo-pituitary-gonadal axis functions properly, but the body is underresponsive to this stimulation. About 90% of cases are idiopathic, or "constitutional." About 5% are due to inadequate function of this endocrine axis, such as GnRH hyposecretion, gonadotropin hyposecretion, or unresponsive gonads. Delayed puberty calls for a thorough assessment with attention to possible chronic illness, eating disorders, drug abuse, endocrine dysfunction, aneuploidies such as Turner or Klinefelter syndrome, and other possibilities. Cases due to hormone hyposecretion can be treated with hormone replacement therapy such as a short course of testosterone treatment. Psychological counseling can help children who are self-conscious about lagging behind their classmates in development.

Aging, Death, and Disease

Because of increasing life expectancy in developed nations, the field of geriatric medicine is developing at a rapid pace. **Geriatric medicine** is an interdisciplinary approach toward managing illness and disability in elderly patients.

Most age-related biological functions are maximized before a person reaches the age of 30. After this, there is a fairly linear decline in any given function. However, the rate of decline is subject to environmental variables—that is, it can be modified by behavior, diet, and other factors. For example, continued exercise helps maintain muscle tone and mass, bone density, and cardiovascular and respiratory function, and avoidance of excessive exposure to sunlight (such as frequent tanning) helps maintain healthy, young-looking skin.

But no matter how hard we try, aging and senescence have some effect on all systems and increase the chances of pathology. An individual's susceptibility to age-related diseases depends on such factors as family history, mental state, nutritional status, activity level, and environmental factors.

As we live longer, the diseases that ultimately cause our demise change. At one time, when life expectancy in developed nations was lower, accidents and infectious diseases were the main causes of death. (In fact, in developing nations, infectious diseases still rank among the leading causes of death.) Now, advanced methods of treatment and prevention have reduced the likelihood of death from infectious disease. Accidents still cause a large number of deaths, but in older people, cancer, heart diseases, and respiratory diseases rank high. Also, as we live longer, we have a higher risk of diseases associated with senescence such as osteoporosis, Alzheimer disease, and Parkinson disease. Even diseases that do not directly cause death can profoundly affect the quality of life. The challenge now facing us is how to provide a high quality of life for people in their "golden years."

Case Study 29 A Teen Pregnancy

Tara is a 16-year-old high school girl who has been sexually involved with her boyfriend Ted for several months. She has kept this a secret so far, but because of her secrecy, she hasn't been able to get birth control pills or be fitted for an IUD. Ted rarely uses a condom, and says he'll "pull out in time." Now, however, Tara has missed two menstrual periods and occasionally feels nauseated in the morning. She buys a home pregnancy testing kit, and the result indicates that she is indeed pregnant. When he gets the news, Ted accuses Tara of getting pregnant by someone else and breaks off their relationship. In desperation, Tara finally decides to confide in her mother.

Tara's mother, Ann, takes her to a gynecologist. A physical examination, urine HCG test, and sonogram all confirm that Tara is pregnant. From the HCG level and the crown-to-rump length of the fetus measured by sonography, the pregnancy is estimated to be 10 weeks along. The fetus and placenta look normal and healthy.

Tara and Ann decide to continue the pregnancy and to do all they can to ensure a healthy baby. Ann, a nurse and a smoker, stops smoking indoors when Tara is at home, and knowing that Tara has sometimes had a few beers at parties with older friends, she cautions her not to drink during the pregnancy for the sake of the baby. Tara has been using Acutane for her complexion, and Ann tells her she'll make an appointment for Tara to see a dermatologist about a safer acne medication. She also tells Tara to lay off the Mountain Dew, iced tea, and other caffeine-containing drinks that she likes so much.

The pregnancy progresses normally. At 20 weeks, Tara has another sonogram and everything appears normal. The heartbeat is clearly visible, and even the sex of the fetus is identifiable—Tara is going to have a baby boy.

Two months later, however, signs of problems begin to appear. Tara has trouble getting her rings off, and she notices that her fingers seem puffy. A few days later, looking in the mirror, she thinks her face is starting to look fat too. She doesn't attach very much importance to this, assuming it to be normal in pregnancy. But not long after that, she notices that her urine is oddly foamy, and she begins missing school because of headaches. When Ann asks how she's feeling, Tara mentions the foamy urine. Ann has seen this in some of her patients in the O.B. ward. Knowing that it might be a sign of preeclampsia, she takes Tara to the gynecologist immediately.

Tara's physical examination and laboratory tests show that her blood pressure is 146/92 and she has proteinuria (a +1 level of albumin in the urine). The doctor advises Tara to stay out of school for a few days, get plenty of bed rest, drink extra water, and come back every 2 days for a checkup. On her second return visit 4 days later, Tara's proteinuria is up to +3, her blood pressure is 154/112, and she has hyperreflexia (slightly exaggerated reflexes). The doctor admits Tara to the hospital immediately, where she is given I.V. saline and then, to prevent convulsions, 4 g of $MgSO_4$ over 15 minutes followed by 2 g of $MgSO_4$ per hour for the next 6 hours. This treatment reduces Tara's hyperreflexia and lowers her blood pressure. She is taken to surgery, and her baby is delivered by cesarian section at a gestational age of 30 weeks and a birth weight of 2.3 kg (5 lb 2 oz). The baby is treated in the neonatal I.C.U. for respiratory distress syndrome.Tara remains in the hospital for 7 days postpartum for observation and stabilization, and the baby is released after 2 weeks. Both eventually go home healthy, but Tara is scheduled for weekly follow-up physicals and urinalysis for the next 6 weeks.

Based on this case study and other information in this chapter, answer the following questions.

1. Specifically where in Tara's kidneys do you think histological damage has occurred as a result of her preeclampsia?
2. Why does Tara's mother take her off Acutane?
3. What do you think might have happened to Tara if the doctor had not admitted her to the hospital, but had released her and told her to come back in another week?
4. Aside from respiratory distress syndrome, suggest some other conditions that Tara's premature baby might exhibit that most neonates do not.
5. When Tara has her first sonogram, do you think a fetal heartbeat is detectable? By the time of her second sonogram, is Tara feeling any fetal movement?

6. Why do you think Tara's baby is not delivered vaginally?
7. Would you classify Acutane as a category A, B, C, D, or X drug? Explain.
8. Why do you think teratogenic drugs have more severe effects on a conceptus prior to 8 weeks of gestation than after 8 weeks?
9. Greg, a 17-year-old male, has not yet developed the same degree of muscle mass or facial hair that his 15-year-old brother Blaine has. Greg's testes are both located in the scrotum, but are smaller than normal. Clinical tests indicate that his serum LH concentration is elevated and his serum testosterone concentration is lower than normal. Based on these findings, Greg is diagnosed with delayed puberty. What treatment could enable Greg to complete pubertal development and develop secondary sex characteristics and normal fertility?
10. Do you think Greg's high LH level is responsible for his low testosterone level, or that his low testosterone level is responsible for his high LH level, or that the two are unrelated? Explain your answer.

Selected Clinical Terms

abruptio placentae Separation of the placenta from the uterine wall before parturition (birth).

delayed puberty The failure of thelarche to begin in a girl by the age of 13 or of testicular enlargement to begin in a boy by the age of 14.

eclampsia Coma, convulsions, or both following preeclampsia.

ectopic pregnancy Implantation and development of a conceptus in any location other than the uterine wall, such as the uterine tube or pelvic cavity.

geriatric medicine The treatment of illness and disability in the elderly.

placenta previa Obstruction of the internal os of the cervical canal by the placenta, or development of the placenta at the margin of the internal os, so that the fetus cannot be delivered before the placenta separates from the uterus.

precocious puberty The beginning of thelarche before the age of 8 in a girl or of testicular enlargement before the age of 9 in a boy.

preeclampsia A syndrome of edema, hypertension, and proteinuria seen usually in the third trimester of pregnancy and usually in primigravidas.

spontaneous abortion Discharge of a conceptus from the uterus before 20 weeks of gestation.